HANDBOOK OF OSCILLOSCOPE WAVEFORM ANALYSIS AND APPLICATIONS

Miles Ritter-Sanders, Jr.

RESTON PUBLISHING COMPANY, INC.
A Prentice-Hall Company
Reston, Virginia

Library of Congress Cataloging in Publication Data

Ritter-Sanders, Miles, Jr.
    Handbook of oscilloscope waveform analysis and applications.

    Includes index.
    1. Cathode ray oscilloscope. I. Title.
TK7878.7.M484   621.3815'48   76-41259
ISBN 0-87909-337-4

© 1977 by Reston Publishing Company, Inc.
A Prentice-Hall Company
Reston, Virginia 22090

All rights reserved. No part of this book may be reproduced in any way, or by any means, without permission in writing from the publisher.

10 9 8 7 6 5 4 3 2 1

Printed in the United States of America

CONTENTS

Preface   vii

One
   Waveform Characteristics and Interrelations   1

      1-1   General Considerations, 1
      1-2   Log and Antilog Waveform Operations, 7
      1-3   Linear and Nonlinear Operations, 11
      1-4   Successive Integration and Differentiation, 16
      1-5   Ideal Versus Real Waveforms, 21
      1-6   Conventionalized Waveforms, 25
      1-7   Basic Waveform-Number Relations, 25
      1-8   Basic Waveform-Circuit Relations, 31
      1-9   Triggered-sweep Oscilloscope Controls, 35

Two
   Audio-Amplifier Malfunctions and Waveform Analysis   39

      2-1   General Considerations, 39
      2-2   Power Bandwidth Measurement, 45

2-3 Music-Power Waveforms, 47
2-4 Evaluation of Waveform Distortion, 51
2-5 Amplifier Sensitivity Measurement and Voltage Gain, 62
2-6 Unbalanced Complementary-Symmetry Output, 63
2-7 Intermodulation Distortion, 63

Three

Stereo-Multiplex Circuit Tests — 67

3-1 General Considerations, 67
3-2 Waveforms in Stereo-Multiplex Generator, 69
3-3 Stereo-Multiplex Waveform Analysis, 83
3-4 Notes on Noise Waveforms, 85
3-5 FM Receiver Alignment, 87

Four

Television Receiver Tests — 91

4-1 General Considerations, 91
4-2 Front-End Tests, 93
4-3 Sync-Pulse Waveform Analysis, 99
4-4 IF Alignment Procedure, 107
4-5 Deflection Current Waveform Checks, 109
4-6 Notes on Conventional Oscilloscope Probes, 113
4-7 Basic Waveshaping Processes, 117
4-8 Intercarrier-sound Section Waveforms, 120
4-9 Three Forms of Sync-buzz Interference, 122
4-10 Antenna and TV Receiver Impedance Matching Tests, 122

Five

Color-TV Receiver Tests — 127

5-1 General Considerations, 127
5-2 Chroma Bandpass Amplifier Tests, 133
5-3 Bandpass-Amplifier Alignment, 135
5-4 Burst-Amplifier Tests, 137
5-5 Subcarrier-Oscillator Checkout, 142
5-6 Chroma Demodulator Checkout, 143
5-7 Specialized Color-TV Test Signals, 151
5-8 SCR Sweep-circuit Waveforms, 153

CONTENTS v

Six
    Industrial-Electronics Circuit Actions and Waveforms   157

    6-1   General Considerations, 157
    6-2   Step-by-step Counting Circuit Waveforms, 161
    6-3   Phase-shifter Waveforms, 161
    6-4   Elimination of Transient in Switched AC Waveform, 165
    6-5   Plasma Oscillation Waveform, 167
    6-6   Clamping Circuit Waveforms, 168
    6-7   Overdamped, Critically Damped, and Underdamped Waveforms, 171
    6-8   Delay Time Measurement With Dual Trace Oscilloscope, 171
    6-9   Inverter Unit Waveform Checks, 173
    6-10   Frequency Converter Waveforms, 174
    6-11   Ferroresonance and Negative Impedance, 175
    6-12   Magnetic Amplifier Waveforms, 176

Appendix 1   RMS Values of Basic Complex Waveforms   179

Appendix 2   Basic Fundamental and Single Harmonic Waveforms   181

Appendix 3   Basic Geometrical Curves   183

Appendix 4   Color-TV Input and Output System Waveforms   186

Appendix 5   I-Q Color Waveform Chart   188

Appendix 6   Vectorscope Waveform Development   191

Appendix 7   Balanced Modulator Waveforms   194

Index   197

PREFACE

The modern oscilloscope is unquestionably a most versatile and informative test instrument. Many engineers and technicians regard it as the most important electronic instrument in the present state of the art. To realize the potential utility of an oscilloscope, the operator must be conversant with the principles of waveform analysis and of application techniques. With the rapid advance in both of these areas within recent years, an urgent need has arisen for a pertinent handbook that is suitable both for self-instruction and for classroom use. This text represents a dedicated effort to meet that need. With the recent shift in academic emphasis in junior colleges, particularly, practical applications have been stressed, and the use of mathematics has been minimized in techniques of waveform analysis. The reader is introduced to the intuitive distinctions between linear and nonlinear circuit actions, and to the resulting characteristics of associated voltage, current, and power waveforms.

Waveform characteristics and interrelations are explained in the first chapter, and some unexpected relations between sine waves and square waves are explained. These relations involve waveform addition, subtraction, and multiplication. Waveform division is described in terms of logarithmic and antilogarithmic processes. Distinctions between ideal and real waveforms are illustrated, and the basic waveform-number relations are explained without recourse to higher mathematics. Audio-amplifier malfunctions and waveform analysis are covered in the second chapter, with attention to the sometimes controversial music-power ratings of amplifiers. Stereo-multiplex tests are discussed, with detailed explanation of vectorscope stereo-separation tests.

Television receiver tests are detailed in the next chapter, with a practical discussion of amplitude and waveshape tolerances. Sync-buzz troubleshooting is covered, and slope-detection waveforms are explained. Color-TV troubleshooting is presented next, including an analysis of specialized test signals. Industrial-electronics circuit actions and waveforms are described in the sixth chapter; the saturable reactor is explained in its basic applications. Ferroresonance, negative impedance, and the basic magnetic-amplifier waveforms are included. The appendixes contain notes on rms values of complex waveforms, basic fundamental and single-harmonic waveforms, and the fundamental geometric curves in their relations to typical electronic-circuit waveforms.

Acknowledgement is made to those who have preceded the author by their development of other books on oscilloscope technology, and to the faculty members, who have made many helpful suggestions and constructive criticisms. This book can be properly described as a team effort, and it is appropriate that the work be dedicated as a teaching tool to the instructors and students of our colleges and technical schools.

Miles Ritter-Sanders, Jr.

1

WAVEFORM CHARACTERISTICS AND INTERRELATIONS

1-1 GENERAL CONSIDERATIONS

A sine wave is an example of an elementary waveform; it is also the basic *steady-state* waveform. Again, a square wave is an example of an elementary *transient* waveform. As shown in Fig. 1–1, both waveforms are characterized in terms of waveshape, amplitude, and frequency. These basic waveforms have important interrelations, some of which may be unexpected. For example, a sine wave can be multiplied by a square wave, and another type of complex waveform will be obtained. If the square wave has no DC component, the product will be a full-rectified sine wave. On the other hand, if the square wave has a single polarity (a DC component of one-half peak value), the product will be a half-rectified sine wave. Thus, full-rectified and half-rectified sine waves may be classified as *product waveforms*.

It is instructive to consider the development of basic product waveforms. Observe that the sine wave and the square wave depicted in Fig. 1–1 are AC waveforms. In other words, these waveforms do not have a DC component. Next, note that the square wave depicted in Fig. 1–2(a) has positive polarity only. That is, this waveform consists of an AC square wave mixed with a DC component of one-half peak value. In turn, the waveform is displayed entirely above the zero-volt axis. On the

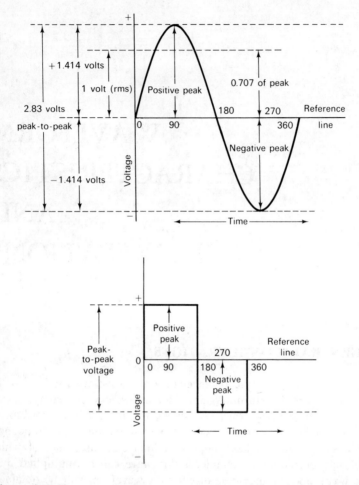

Figure 1-1 Sine waves and square waves have unexpected interrelations.

other hand, an AC square wave is depicted in Fig. 1-2(b); this waveform is centered on the zero-volt axis. Both varieties of square waveforms find extensive application in analog computers and in various types of wave-shaping equipment.

Waveforms may be added, subtracted, multiplied, or divided. For example, if a summing probe such as that depicted in Fig. 1-3 is used with an oscilloscope, the sum of two input waveforms will be displayed on the screen. As pictured in Fig. 1-4, the addition of a sine wave and a square wave produces a square wave that has a boosted fundamental. Of course, both input waveforms have the same frequency (repetition

SECT. 1-1 / GENERAL CONSIDERATIONS

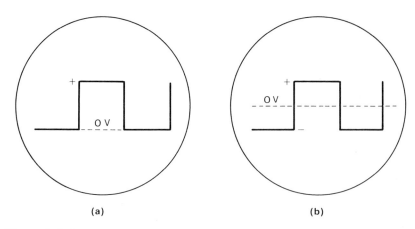

Figure 1-2 Square waves with and without DC component: (a) with DC component of one-half peak value; (b) without DC component.

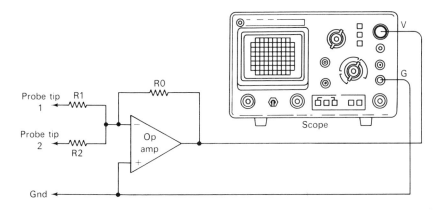

Figure 1-3 Basic summing probe.

rate) in this example. Moreover, the input sine wave has the same phase as the fundamental component of the input square wave. In the event that the input sine wave were shifted 180 deg in phase, the output from the summing probe would be the difference between the two waveforms. In other words, the sine wave would be subtracted from the square wave, and the output waveform would be a square wave with a concave top and bottom excursion. This waveform operation is illustrated in Fig. 1-5.

Observe that the operation shown in Fig. 1-5 is an aspect of waveform filtering. This filtering action may be partial or complete, depending on the relative amplitudes of the input waveforms. Next, it is

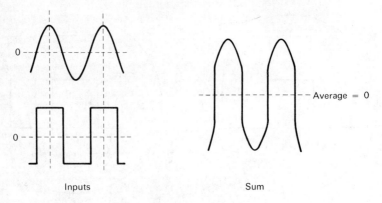

Figure 1–4 Addition of sine wave to square wave forms a sum that is a square wave with boosted fundamental.

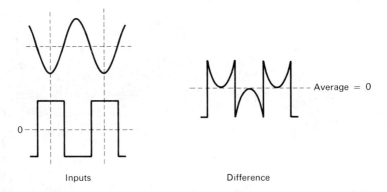

Figure 1–5 Subtraction of sine wave from square wave forms a difference that is a square wave minus its fundamental component.

instructive to consider the multiplication of a pair of waveforms. With reference to Fig. 1–6, three operational amplifiers may be interconnected as shown to provide a versatile waveform multiplier. Briefly, it may be noted that when amplifier stages operate in cascade, the gain of the first stage is multiplied by the gain of the second stage. As pictured in Fig. 1–7, when a sine wave is multiplied by a square wave, the output is a product waveform that has the shape of a full-rectified sine wave. Note in this example that neither the sine-wave input nor the square-wave input has a DC component. Also, the sine-wave input has the same phase as

SECT. 1-1 / GENERAL CONSIDERATIONS

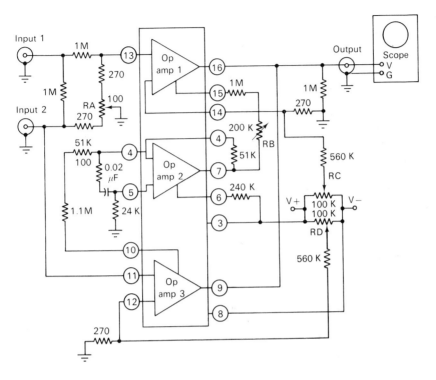

Figure 1–6 A versatile op-amp waveform multiplier arrangement.

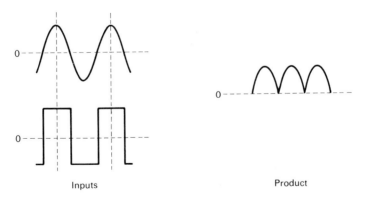

Figure 1–7 Multiplication of a sine wave by a square wave without a DC component forms a product that is a full-rectified sine wave.

the fundamental component of the square-wave input, and both input waveforms have the same frequency (repetition rate).

Observe that if the sine-wave input were 180 deg out of phase with respect to that exemplified in Fig. 1–7, the output would be a product waveform with an opposite polarity. In other words, the output waveform depicted in Fig. 1–7 has positive polarity. On the other hand, if the sine-wave input were shifted 180 deg in phase, the output waveform would have negative polarity. Note also that the output waveform has a DC component in either case, in spite of the fact that neither of the input waveforms has a DC component. This DC component derives from the circumstance that the product of two negative quantities is a positive quantity, whereas the product of two positive quantities is again a positive quantity. With reference to Fig. 1–8, if the square-wave input waveform has a DC component that is equal to one-half peak value, it is evident that the square-wave voltage is zero during the negative excursion of the input sine wave. In turn, the output product waveform has the shape of a half-rectified sine wave.

Of course, the same waveform may be applied to both inputs of an op-amp multiplier. In such a case, the waveform is multiplied by itself; in other words, the waveform is squared. If a sine wave is squared, as depicted in Fig. 1–9, the output is a product waveform that is a sine wave which has twice the frequency of the input waveforms. In other words, this operation is basically frequency-doubling. Note also that the output waveform has a DC component that makes the excursion entirely positive. Again, this introduction of a DC component results from the circumstance that the product of two positive quantities is a positive quantity, and the product of two negative quantities is also a positive

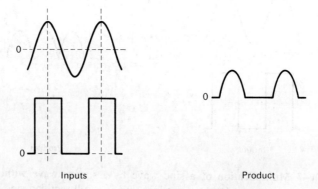

Inputs   Product

Figure 1–8 Multiplication of sine wave by square wave with DC component forms a product that is a half-rectified sine wave.

SECT. 1-2 / LOG AND ANTILOG WAVEFORM OPERATIONS   7

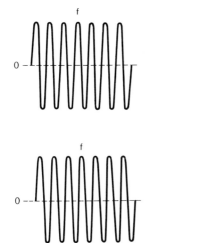
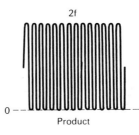

Figure 1–9 Multiplication of a sine wave by itself forms a product that is a double-frequency sine wave (sine-squared waveform).

quantity. Observe that if one of the input waveforms were shifted in phase by 180 deg, the output waveform would remain the same, except that it would then have a negative DC component.

Two successive operations may be employed to obtain a parabolic waveform from a square wave. For example, an op-amp integrator may be utilized as depicted in Fig. 1–10 to form a triangular wave from a square wave. Note in passing that a triangular waveform is the true mathematical integral of a square waveform. In turn, if a triangular wave is applied to both inputs of an op-amp multiplier, the triangular waveform becomes squared, and the output is a product waveform that has a parabolic waveshape, as exemplified in Fig. 1–11. These operations are commonly employed in function generators to produce various complex output waveforms. A typical function generator provides sine, square, sawtooth, parabolic, and pulse output waveforms.

1-2 LOG AND ANTILOG WAVEFORM OPERATIONS

Multiplication and division of waveforms is often accomplished by means of op-amp logarithmic and antilogarithmic arrangements. Involution (rooting) and evolution (raising to a power) are also generally

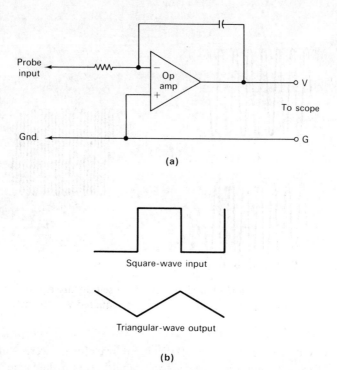

Figure 1-10 Op-amp integrator arrangement: (a) configuration; (b) corresponding input and output waveforms.

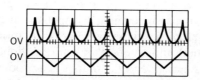

Figure 1-11 A parabolic waveform results from squaring a triangular waveform.

accomplished by log arrangements. These basic configurations are depicted in Fig. 1-12. Semiconductor junctions are employed to develop the log and antilog functions. In the basic log amplifier configuration, the base-emitter junction of a bipolar transistor operates in the negative-feedback loop to provide an output voltage that is proportional to the log of the input voltage. Conversely, in the basic antilog amplifier configuration, the anode-cathode junction of a semiconductor diode operates in the input circuit to provide an output voltage that is proportional to

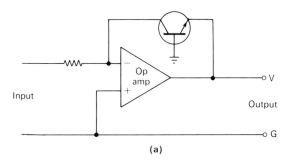

(a)

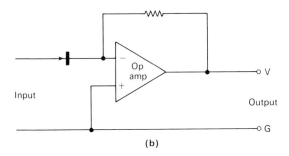

(b)

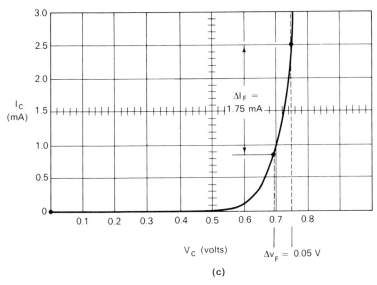

(c)

Figure 1-12 Op-amp logarithmic and antilogarithmic arrangements: (a) basic log amplifier; (b) basic antilog amplifier; (c) typical semiconductor junction characteristic.

the antilog of the input voltage. As shown in Fig. 1–12(c), the forward characteristic of a semiconductor diode junction has an essentially logarithmic form.

Two quantities can be multiplied by adding their logarithms and then taking the antilogarithm of this sum. Accordingly, a sine wave can be multiplied by a square wave as follows:

1. Pass the sine wave through a logarithmic amplifier.
2. Pass the square wave through another logarithmic amplifier.
3. Apply the amplifier outputs to a summer such as that shown in Fig. 1–3.
4. Pass the summer output through an antilog amplifier.
5. An output waveform that is the product of the sine wave and of the square wave will be obtained from the antilog amplifier.

One quantity can be divided by another quantity if their logarithms are subtracted, and the difference applied to an antilog amplifier. A basic op-amp subtracter arrangement is shown in Fig. 1–13. Note that if one waveform is applied to the noninverting input (input 2), and another waveform is applied to the inverting input (input 1), the output waveform will be the difference between the two input waveforms. In other words, the voltage applied to input 1 is subtracted from the voltage applied to input 2. Accordingly, a sine wave can be subtracted from a triangular wave as follows:

1. Pass the sine wave through a logarithmic amplifier.
2. Pass the square wave through a logarithmic amplifier.
3. Apply the amplifier outputs to an op-amp subtracter.

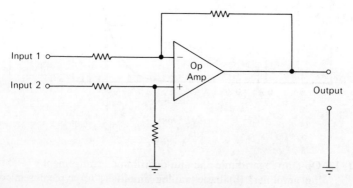

Figure 1–13 Basic op-amp subtractor arrangement.

SECT. 1-3 / LINEAR AND NONLINEAR OPERATIONS   11

4. Pass the subtracter output through an antilog amplifier.
5. An output waveform that is the quotient of the triangular wave and the sine wave will be obtained from the antilog amplifier.

Next, consider the process of raising a waveform to a given power. For example, it may be desired to obtain the cube of a sine wave. This can be accomplished by multiplying the log of the sine wave by 3, and passing the product through an antilog amplifier. Thus, the third power of a sine wave is obtained as follows:

1. Pass the sine wave through a logarithmic amplifier.
2. Apply the output from the log amplifier to an op-amp multiplier.
3. Also apply three units of DC voltage to the multiplier.
4. Pass the multiplier output through an antilog amplifier.
5. An output waveform that is the third power of the input sine wave will be obtained from the antilog amplifier.

Again, consider the process of extracting a given root of a specified waveform. As an illustration, it may be desired to extract the cube root of a sine wave. This can be accomplished by dividing the log of the sine wave by 3, and passing the quotient through an antilog amplifier. Thus, the cube root of a sine wave is obtained as follows:

1. Pass the sine wave through a logarithmic amplifier.
2. Apply the output from the log amplifier to an op-amp divider (Fig. 1–14).
3. Also apply three units of DC voltage to the divider.
4. Pass the divider output through an antilog amplifier.
5. An output waveform that is the cube root of the input sine wave will be obtained from the antilog amplifier.

1-3 LINEAR AND NONLINEAR OPERATIONS

Waveforms may be processed by linear networks or by nonlinear networks. A linear network is specifiable by linear differential equations with time as the independent variable, whereas a nonlinear network is not specifiable by such equations. From a practical viewpoint, a linear operation introduces no new frequencies into a waveform, whereas a nonlinear

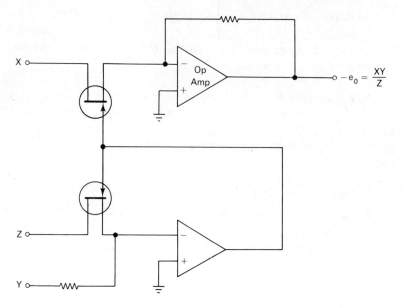

Figure 1–14 Basic op-amp divider arrangement.

operation introduces new frequencies into the output waveform. For example, consider the examples of linear and nonlinear mixing depicted in Fig. 1–15. When a low-frequency sine wave is linearly mixed with a higher-frequency sine wave, the frequency spectrum of the resulting waveform contains no frequencies other than the input frequencies. On the other hand, when a low-frequency sine wave amplitude-modulates a higher-frequency sine wave, the frequency spectrum of the resulting waveform contains frequencies other than the input frequencies. In this particular example, the new frequencies are called *sideband frequencies*. Thus, amplitude modulation is a nonlinear process.

Capacitance and resistance are ordinarily linear circuit elements. In turn, RC circuitry does not introduce new frequencies into the output waveform. Air-core inductors are essentially linear. On the other hand, an iron-core inductor may exhibit significant nonlinearity owing to core-saturation action. As another practical example, a high-fidelity audio transformer is a virtually linear component, whereas a power-supply transformer usually exhibits appreciable nonlinearity. When a transformer processes an AC waveform without any DC component present in either the primary or secondary windings, core saturation tends to compress both the positive and the negative peaks of the output waveform. This mode of distortion is chiefly third-harmonic, and the output waveform is said to contain an *iron third harmonic*.

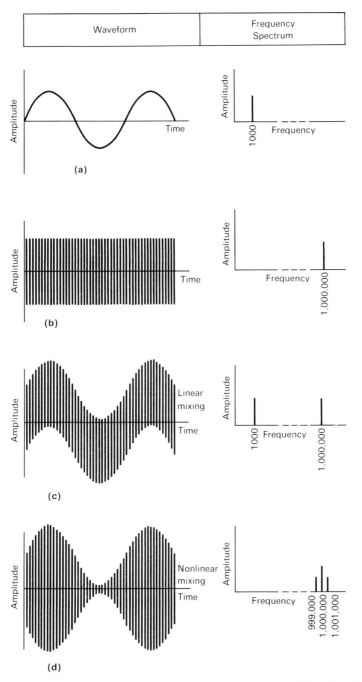

Figure 1-15 Examples of linear mixing versus nonlinear mixing (amplitude modulation): (a) sine wave; (b) higher-frequency sine wave; (c) mixing of (a) and (b); (d) modulation of (b) by (a).

A semiconductor diode may be operated as an essentially linear device, or as a highly nonlinear device. With reference to Fig. 1–12, the interval from 0.6 to 0.7 volt on the voltage-current (E-I) characteristic of a silicon diode is a highly nonlinear region. On the other hand, the interval from 0.7 to 0.8 volt is somewhat nonlinear. Again, the interval from 0.8 to 0.9 volt is essentially linear. When operated as a modulator or as a demodulator device, nonlinear action is required; in turn, the interval from 0.6 to 0.7 volt is suitable for these applications. On the other hand, if the diode were to be employed as a voltage-level shifter, the interval from 0.7 to 0.8 volt would be more appropriate. If greater linearity were desirable, the diode could be operated over its 0.8-to-0.9-volt interval.

Similarly, a transistor may be operated as an essentially linear device, or as a highly nonlinear device. For example, if a transistor is employed as a demodulator, it will be operated over a highly curved interval of its base-emitter characteristic. However, if the transistor is to be utilized as a class-A amplifier, it will be operated with sufficient forward bias that the input signal excursion occurs over an essentially linear interval of its base-emitter characteristic. When minimum waveform distortion is required, as in a high-fidelity amplifier, substantial negative feedback is generally used to obtain optimum linearity. Nonlinear devices are sometimes operated to exploit opposing nonlinear characteristics, thereby providing a linear transfer characteristic. For example, an amplifier transistor may be biased into a nonlinear interval of operation so that an applied exponential waveform is amplified and changed into a linear sawtooth waveform. In other words, the stage is operated as a wave-shaping amplifier.

Some waveforms have even harmonics only. Other waveforms have odd harmonics only. Still other waveforms have both even and odd harmonics, as exemplified in Fig. 1–16. If an even-harmonic waveform is passed through a linear circuit, an even-harmonic output waveform is obtained. Or, if an odd-harmonic waveform is passed through a linear circuit, an odd-harmonic output waveform is obtained. Again, if an even-and-odd harmonic waveform is passed through a linear circuit, an even-and-odd harmonic output waveform is obtained. As an illustration, consider the differentiated waveforms shown in Fig. 1–17. Since capacitors and resistors are linear circuit elements, a differentiating circuit is a linear arrangement. In turn, the output waveforms will contain no new harmonics with respect to the input waveform. Of course, the relative amplitudes of the harmonics are changed by the differentiation process. Relative phases of the harmonics are also changed. However, differentiation is a linear operation, inasmuch as no new harmonics are introduced.

SECT. 1-3 / LINEAR AND NONLINEAR OPERATIONS   15

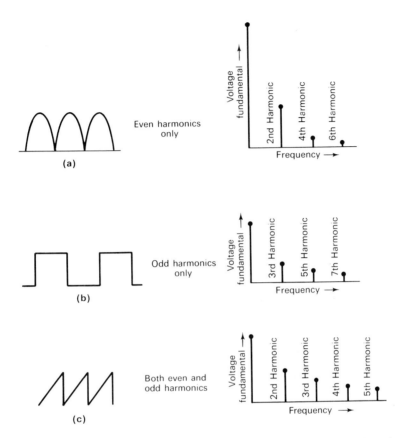

Figure 1-16 Basic harmonic relations in standard waveforms: (a) full-rectified sine wave; (b) square wave; (c) sawtooth wave.

An op amp may be operated either as a linear or as a nonlinear system. For example, the op amp in Fig. 1-10 operates as a linear device in the integrator arrangement. On the other hand, consider the action of an op-amp multiplier. With reference to Fig. 1-7, observe that one input waveform has a fundamental frequency only, and the other input waveform has odd harmonics only. However, the output waveform has a fundamental frequency and even harmonics only. In other words, the op amp operates as a nonlinear device in this example. An op-amp adder or subtracter operates as a linear device. Note that a subtracter may be operated as a filter to remove the fundamental frequency from a complex input waveform. In turn, the output waveform has a missing

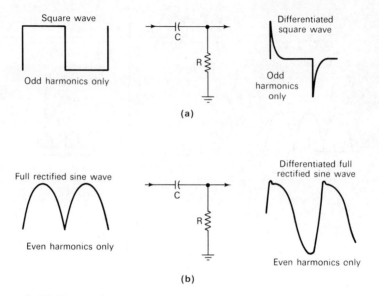

Figure 1-17 Harmonic frequencies remain unchanged in a linear system: (a) differentiation of a square wave; (b) differentiation of a full-rectified sine wave.

fundamental frequency. This is a linear operation, inasmuch as the output waveform contains no new frequencies with respect to the input waveforms.

1-4 SUCCESSIVE INTEGRATION AND DIFFERENTIATION

Electronic equipment may employ successive integration, as in the vertical-sync section of a television receiver. A single RC section integrates a square wave or pulse waveform into an exponential waveform. With reference to Fig. 1-18, the rise time of this exponential waveform is a function of the **integrator time constant**. That is, the output waveform will rise to approximately 63 percent of its ultimate amplitude in one time constant. An exponential waveform follows the natural law of growth and decay. Note that the great majority of complex waveforms encountered in television receiver circuitry are modified exponential waveforms. As shown in Fig. 1-19, when a square wave input is applied

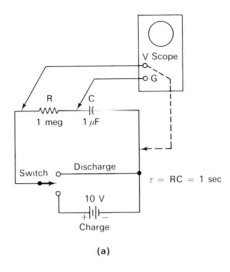

(a)

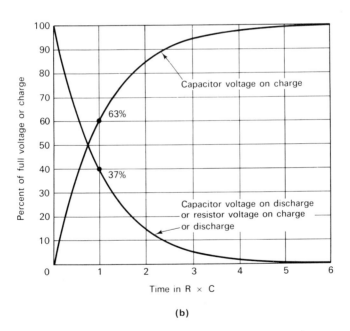

(b)

Figure 1–18 Basic integrating and differentiating circuit action: (a) RC circuit with a time constant of one second; (b) universal time-constant chart for series RC circuit.

18   CHAP. 1 / WAVEFORM CHARACTERISTICS

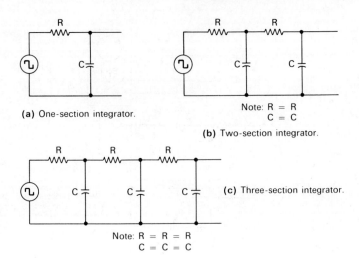

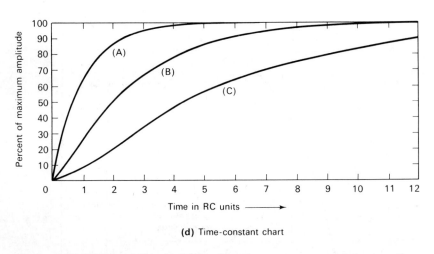

(d) Time-constant chart

Figure 1-19 Universal time-constant chart for one-, two-, and three-section integrator circuits: (a) one-section integrator; (b) two-section integrator; (c) three-section integrator; (d) time-constant chart.

to a two-section or a three-section integrator, a modified exponential output waveform is obtained. Each successive RC section slows down the rise of the output waveform, and also modifies its waveshape. It is instructive to consider the comparative responses of multisection inte-

SECT. 1-4 / SUCCESSIVE INTEGRATION AND DIFFERENTIATION   19

grator circuits to an applied pulse, as exemplified in Fig. 1–20. In this example, the input pulse has an exponential decay. As the pulse is processed by successive integrating circuit sections, the output pulse becomes attenuated, delayed, and modified in waveshape. Observe also that the output waveform becomes more nearly symmetrical as more integrating sections are included.

Next, consider the square-wave response of a two-section differentiating circuit with and without device isolation, as shown in Fig. 1–21. In each case, there is an undershoot following the decay interval; however, this undershoot is increased when device isolation is employed. Decay is faster when direct coupling is utilized. Note that undershoot occurs because cascaded RC differentiating circuits have an inherent equivalent inductive response that develops a counter electromotive force. It follows from previous discussion that a differentiating circuit is a simple form of high-pass filter. Similarly, an integrating circuit is a simple form of low-pass filter. When suitable RC values are used in a differentiating circuit followed by an integrating circuit, a simple form of bandpass filter is formed. This arrangement is typically employed in television vertical-sync networks. As noted above for cascaded differentiating circuits, an RC bandpass filter arrangement introduces undershoot into the output waveform.

An RC-coupled amplifier has effective shunt capacitance and series resistance. In turn, a two-stage RC amplifier has a basic equivalent circuit, as depicted in Fig. 1–22. Device isolation operates to modify the output

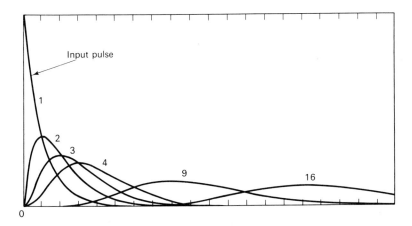

Figure 1–20 Responses produced by successive integrating circuit sections to an applied pulse.

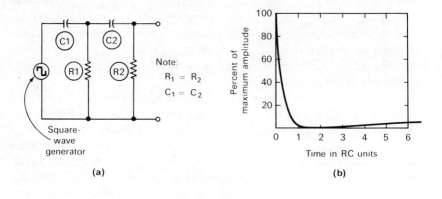

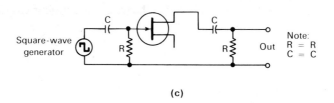

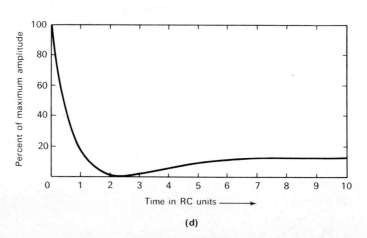

Figure 1–21 Responses of two-section differentiating circuits: (a) direct-coupled two-section differentiating circuit; (b) response of (a) to square-wave input; (c) two-section differentiating circuit with device isolation; (d) response of (c) to square-wave input.

SECT. 1-5 / IDEAL VERSUS REAL WAVEFORMS

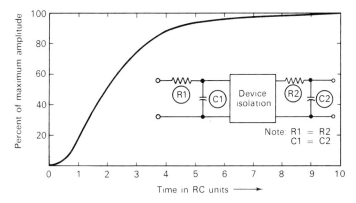

Figure 1-22 Square-wave response of two-section integrating circuit with device isolation.

waveshape to some extent. Comparison of the response in Fig. 1-22 with the response for the two-section integrator in Fig. 1-19 shows that device isolation results in somewhat greater delay at the outset, although 90 percent of the total rise is completed in practically the same period of time. In turn, device isolation results in a somewhat steeper wavefront. Two time constants are required for the output waveforms to rise to 50 percent of their final amplitude, whether or not device isolation is utilized. Thus, device isolation has a greater effect on waveform processing in cascaded differentiating circuits than in cascaded integrating circuits. Fig. 1-23 summarizes the waveform characteristics and interrelations that have been discussed, and their relations to topics covered in following chapters.

1-5 IDEAL VERSUS REAL WAVEFORMS

All waveforms have an ideal aspect and a real aspect. Ideal waveforms are described by various mathematical relations. Real waveforms represent practical physical approximations to ideal waveforms. As an illustration, Fig. 1-24 shows the television synchronizing waveform specified by the FCC. On the other hand, Fig. 1-25 exemplifies typical real waveforms corresponding to the foregoing ideal waveforms. In Fig. 1-24, only positive-going sync (black-level sync) is pictured. In practice, however, the television synchronizing waveform is often inverted, as

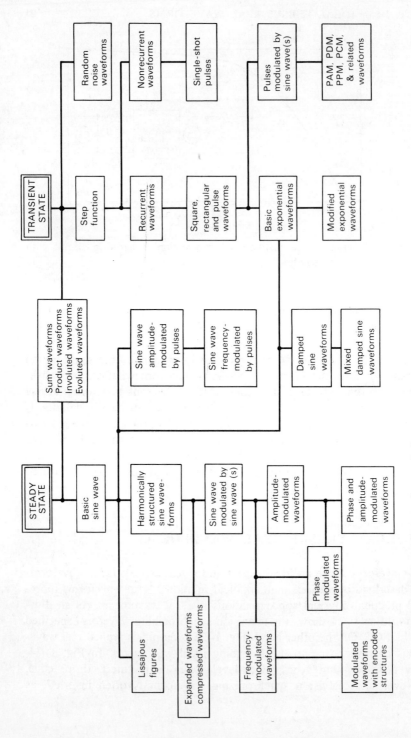

Figure 1-23 Basic waveform relations.

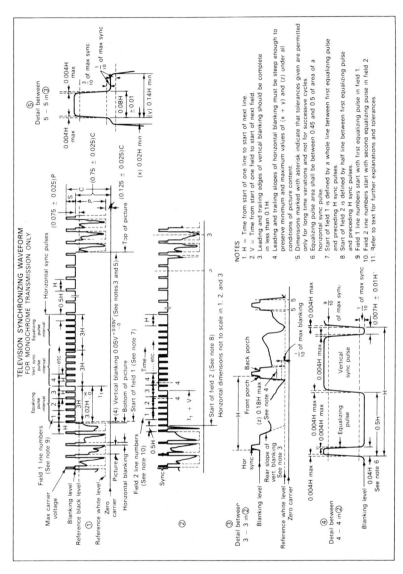

Figure 1-24 FCC specifications for the television synchronizing waveform.

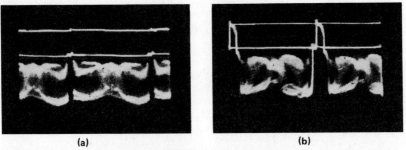

Figure 1-25 Example of real waveforms corresponding to ideal waveforms shown in Fig. 1-24: (a) displayed on 30-Hz deflection; (b) displayed on 7,875-Hz deflection.

exemplified in Fig. 1-26. Observe in Figs. 1-25 and 1-26 that real waveforms depart from ideal waveforms in respect to rise time. In other words, an ideal sync waveform has zero rise time, whereas a real sync waveform has a finite rise time. Also, an ideal sync waveform has perfectly square corners and flat tops, whereas a real sync waveform has corners that are sometimes rounded, or sometimes distorted by overshoot, and tops that are rounded and/or sloped.

Again, a comparison of ideal and real digital pulses is shown in Fig. 1-27. Note that the ideal pulses are perfectly rectangular with flat tops and zero rise time. On the other hand, the oscilloscope pulse display consists of waveforms that are not rectangular, that have rounded tops, and that have finite rise time. Also, observe that there is a fall-time irregularity in the pulse waveforms displayed on the oscilloscope screen (Fig. 1-27). Minor baseline irregularities are also visible. Although real waveforms have various degrees of approach to ideal waveforms, it is impossible to produce an ideal waveform by electronic circuit action. Therefore, real waveforms that have satisfactory correspondence to their

Figure 1-26 Another example of real waveforms corresponding to ideal waveforms shown in Fig. 1-24: (a) displayed on 30-Hz deflection; (b) displayed on 7,875-Hz deflection.

SECT. 1-7 / BASIC WAVEFORM-NUMBER RELATIONS

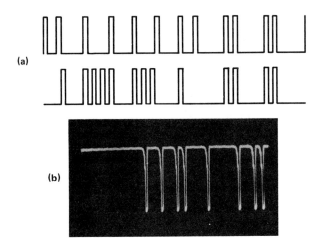

Figure 1-27 Comparison of ideal and real digital pulse waveforms: (a) examples of ideal digital pulses; (b) display of pulses for a digital computer.

ideal counterparts are specified in equipment operation. These specified waveforms are key troubleshooting waveforms that are checked by the technician in the event of equipment malfunction.

1-6 CONVENTIONALIZED WAVEFORMS

Waveforms are sometimes conventionalized, as exemplified in Fig. 1-28. A color burst consists of 8 or 9 cycles of a 3.58-MHz sine wave. Note that if the oscilloscope is synchronized with respect to the horizontal-sync pulse, as in Fig. 1-28(a), the burst waveform is likely to appear blurred, with the individual sine-wave excursions displayed as an area of gray shading. On the other hand, if an oscilloscope is synchronized with respect to the burst waveform, the individual 3.58-MHz cycles will be displayed on the scope screen. In the first conventionalization shown in Fig. 1-28(b), the sine-wave cycles are simulated and are represented as a semi-square waveform. Again, in Fig. 1-28(c), the sine-wave cycles are represented as a triangular waveform. Finally, in Fig. 1-28(d), the sine-wave cycles are represented as a square waveform. Thus, the troubleshooter does not expect to observe either ideal waveforms or conventionalized waveforms.

1-7 BASIC WAVEFORM-NUMBER RELATIONS

It follows from the foregoing discussion that there are basic relations between waveforms and numbers. For example, the amplitude of a

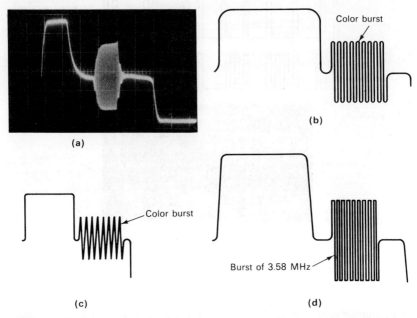

Figure 1-28 Real color-burst display and conventionalizations: (a) display of color burst on oscilloscope screen; (b) conventionalization with sine-wave approximation; (c) another sine-wave approximation; (d) still another sine-wave approximation.

waveform is described by stating that it has a peak-to-peak voltage of 10 volts. Again, the frequency of a waveform is described by stating that its repetition rate is 15,750 Hz. Or the phase of a waveform is described as 30 deg with respect to a reference waveform. These are examples of positive numbers, and their relations to waveforms are readily understood. In addition to positive numbers, negative numbers are also used to describe waveform characteristics. As an illustration, we state that the positive-peak voltage of a sine wave is 12 volts, and that its negative-peak voltage is -12 volts. A negative voltage value is not as readily understood as a positive voltage value. In other words, we know the meaning of the statement "2 oscilloscopes" and of the statement "2 volts." On the other hand, a statement such as "-2 oscilloscopes" is meaningless, although the statement "-2 volts" is readily understandable. This is just another way of saying that we cannot count objects that do not exist, and that we speak of "-2 volts" in a special sense. In this example, we define negative voltage as extending downward below the zero level on the oscilloscope screen. Note carefully that a negative voltage is just as real as a positive voltage. These waveform-number relations are pictured in Fig. 1-29.

SECT. 1-7 / BASIC WAVEFORM-NUMBER RELATIONS

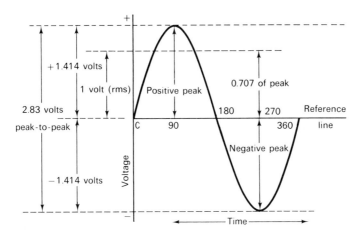

Figure 1–29 Negative and positive voltages in a sine wave.

Observe in the diagram of Fig. 1–29 that the peak-to-peak voltage of the waveform is indicated as 2.83 volts. In other words, it would appear that when +1.414 volts is added to −1.414 volts, the total should be zero, instead of 2.83 volts. Of course, this apparent contradiction occurs because the positive and negative signs are being used only to indicate directions with respect to the reference line, or zero level. As explained above, the negative excursion of the sine wave is just as real as its positive excursion. Therefore, we understand that we must disregard the plus and minus signs when we add the positive-peak voltage to the negative-peak voltage to find the peak-to-peak voltage of the waveform. Technically, we consider the *absolute values* of the positive-peak and the negative-peak voltages when we calculate the peak-to-peak voltage value. An absolute value of a voltage is simply its numerical value, without regard to positive or negative polarities.

Somewhat the same principles are involved in waveform phase relations. As an illustration, consider the phase relations of the R-Y, B-Y, and −(G-Y) waveforms depicted in Fig. 1–30. In this example, the three phases are referenced to burst phase. Thus, the R-Y waveform goes through its positive-peak value 90 deg after the burst waveform, or the R-Y waveform has a phase of 90 deg with respect to burst. Again, the B-Y waveform goes through its positive-peak value 180 deg after the burst waveform, or the B-Y waveform has a phase of 180 deg with respect to burst. Finally, the −(G-Y) waveform goes through its positive-peak value 120 deg after the burst waveform, or the −(G-Y) waveform has a phase of 120 deg with respect to burst. Note, first, that a −(G-Y) waveform is 180 deg out of phase with a G-Y waveform. This

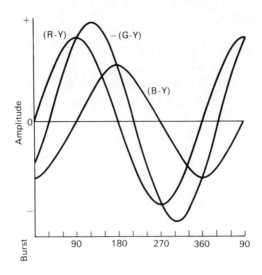

Figure 1–30 Phase relations of R-Y, B-Y, and −(G-Y) waveforms.

relationship is illustrated in Fig. 1–31. Observe that the +(G-Y) and −(G-Y) designations have nothing to do with positive excursions and negative excursions. In other words, these two waveforms are simply defined in phase as illustrated.

Next, with reference to Fig. 1–30, consider the phase relation of the R-Y waveform with respect to the B-Y waveform. When the B-Y waveform is taken as the reference waveform, the phase of the R-Y waveform is then −90 deg. In turn, it is evident that negative and positive phase values are merely defined, and that they have no necessary significance with regard to addition or subtraction of phase angles. Instead of assigning positive and negative phase angles, oscilloscope technicians often state that one waveform *leads* or *lags* another waveform. For example, consider the voltage and current waveforms shown in Fig. 1–32. These are the waveforms that are displayed by a practically pure

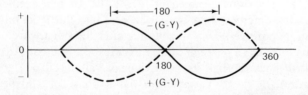

Figure 1–31 Illustration of G-Y and −(G-Y) waveforms.

SECT. 1-7 / BASIC WAVEFORM-NUMBER RELATIONS

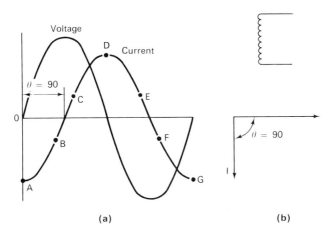

Figure 1-32 Current waveform lags voltage waveform in an inductor: (a) oscilloscope display; (b) vector diagram.

inductor. Observe that the current waveform lags the voltage waveform, and that the voltage waveform leads the current waveform. On the other hand, with reference to Fig. 1-33, a capacitor has a current waveform that leads its voltage waveform by 90 deg, or its voltage waveform lags its current waveform by 90 deg.

Still another basic waveform-number relation is encountered in a power waveform for a capacitor, as shown in Fig. 1-34. Power is equal to the product of voltage and current. In the case of a capacitor, the

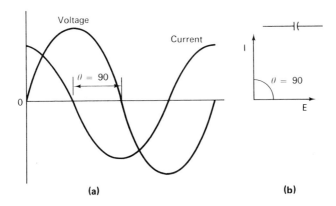

Figure 1-33 Current waveform leads the voltage waveform in a capacitor: (a) oscilloscope display; (b) vector diagram.

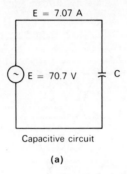

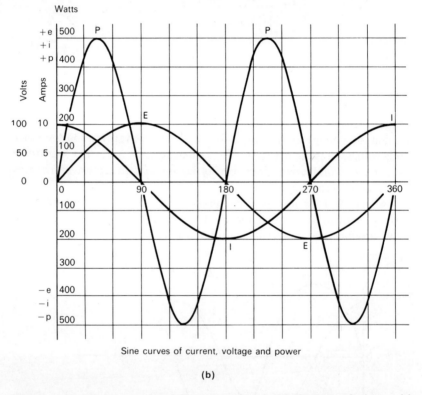

Figure 1–34 Power waveform for a capacitor is 90 deg out of phase with the voltage waveform and with the current waveform.

voltage and current waveforms are 90 deg out of phase with each other. Accordingly, the voltage and current have the same polarity for half of the time, and have different polarities for half of the time. In turn, the product of voltage and current is positive for half of the time, and is negative for half of the time. It follows that the average value of the power waveform is zero. Practically, this means that the power is merely surging back and forth in the circuit, first in one direction, and then in the other direction. Note carefully that the source supplies no power to the capacitor, and that no power is dissipated in the circuit. For this reason, the power in a capacitor is called *reactive power, reactive volt-amperes, imaginary power,* or *wattless power.*

By way of comparison, the power in a resistor is called *real power,* or *true power.* Real power is measured in watts, whereas reactive power is measured in VARs. Note that the term VAR is an abbreviation of volt-amperes-reactive. Just as the power in a capacitor is imaginary power, the current into and out of a capacitor is called *imaginary current.* This means that the current does no work. Imaginary current is designated by imaginary numbers, whereas real current is designated by real numbers. For example, 2 mA denotes real current, whereas $j2$ mA denotes imaginary current. Note that the term j symbolizes $\sqrt{-1}$, the basic imaginary number. Of course, an imaginary current such as $j2$ mA produces a real pattern on an oscilloscope screen. The term "imaginary" simply means that the capacitor current is 90 deg out of phase with the capacitor voltage.

Just as a capacitor draws an imaginary or reactive current, so does an ideal inductor draw an imaginary or reactive current. However, there is a basic distinction between capacitive currents and inductive currents, in that they are 180 deg out of phase with each other. To distinguish between capacitive current and inductive current, technicians write $+j2$ mA for an inductive current, and write $-j2$ mA for a capacitive current. This terminology denotes that the inductive current lags a resistive current by 90 deg, and that the capacitive current leads a resistive current by 90 deg. Of course, a resistive current is always in phase with the applied voltage—its phase angle is zero deg. To summarize briefly, various waveforms will be technically described in terms of positive and negative real numbers, and in terms of positive and negative imaginary numbers. It is also instructive to note that if a real sine wave is added to an imaginary sine wave of the same frequency, a third sine wave is formed, which is represented by a complex number. For example, $2 + j3$ is a complex number that has a real part equal to 2 and an imaginary part equal to 3.

1-8 BASIC WAVEFORM-CIRCUIT RELATIONS

Waveform analysis is facilitated by recognizing various waveform-circuit relations. From the most general viewpoint, circuits are classified as linear or nonlinear arrangements. Thus, a class-A amplifier is a linear configuration, whereas a rectifier circuit is a nonlinear configuration. The term *linear* is defined as an input/output relationship in which the output varies in direct proportion to the input. On the other hand, the term *nonlinear* is defined as an input/output relationship in which the output does not vary in direct proportion to the input. An attenuator is comparable to an amplifier, except that the former reduces the signal level, whereas the latter increases the signal level. Although conventional attenuators are ordinarily linear arrangements, they may become nonlinear under fault conditions. For example, Fig. 1–35 shows the configuration for a simple compensated attenuator. If C403 becomes open-circuited, a complex waveform will become distorted in passage through the attenuator, and its input/output relationship will become nonlinear. This system is nevertheless linear with respect to any individual frequency component (fundamental or harmonic) of the input complex waveform.

A delay line is another example of an ordinarily linear configuration. Its function is to provide a finite transit time for a waveform, so that its output is delayed with respect to its input. Thus, Fig. 1–36 exemplifies

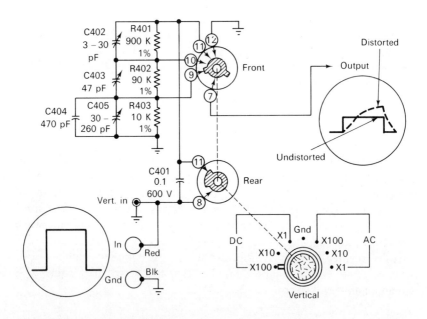

Figure 1–35 Vertical-attenuator configuration for a service-type oscilloscope. (*Courtesy of* Heath Co.)

SECT. 1-8 / BASIC WAVEFORM-CIRCUIT RELATIONS   33

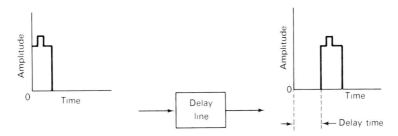

Figure 1-36 Ideal delay-line action.

ideal delay-line operation. Observe that the output pulse is a precise replica of the input pulse, but that its appearance at the output of the delay line occurs substantially after the application of the input pulse. In practice, the output waveform is not necessarily a precise replica of the input pulse. In this respect, a practical delay line departs more or less from a linear system. However, the essential requirement is that a practical delay line should introduce only minor nonlinearity. For example, consider the relations shown in Fig. 1-37. A Y-signal delay line in a color-TV receiver normally introduces a delay of almost 1 microsecond in signal passage. Observe that, although the output waveform is not an exact replica of the input waveform, the distortion that occurs is minor, and a technician will consider that circuit action is acceptable in this example.

In many systems, linear or nonlinear circuit action depends upon the type of input signal that is applied. As an illustration, consider the simple bridged equalizer depicted in Fig. 1-38. With regard to the response of the equalizer alone, it is evident that the RCL circuit has a linear response to a single sine-wave input. On the other hand, this RCL circuit will have a nonlinear response to a square-wave input. Next, with regard to the response of the equalizer and line combination, a uniform frequency response is obtained. Accordingly, the system will have a linear response to either a sine-wave or to a square-wave input. Note that a linear phase characteristic has been assumed for the system; if the phase characteristic happens to be substantially nonlinear, a square-wave signal will be distorted in passage. Elaborate equalizer arrangements are designed to provide both uniform frequency response and a linear phase characteristic for the system.

Next, consider the diode-detection arrangement depicted in Fig. 1-39. This may be regarded as a linear system or as a nonlinear system,

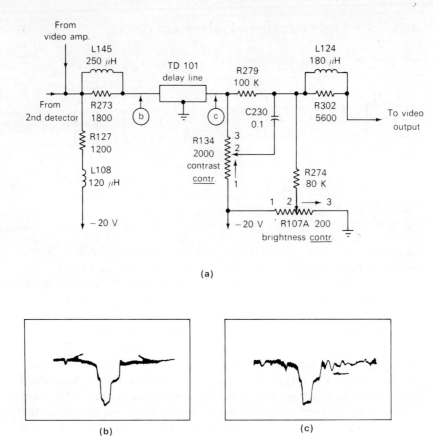

Figure 1–37 Practical delay-line action: (a) configuration; (b) input waveform; (c) output waveform.

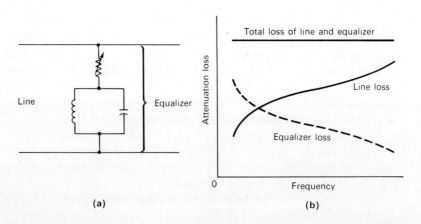

Figure 1–38 A simple bridged equalizer: (a) configuration; (b) frequency response.

SECT. 1-9 / TRIGGERED-SWEEP OSCILLOSCOPE CONTROLS

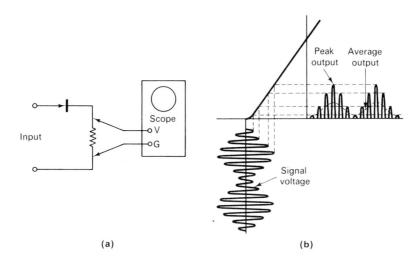

Figure 1-39 Basic diode detector response: (a) configuration; (b) input-output waveforms.

depending upon the circuit-action aspect that is chosen. Since the diode eliminates the negative half-cycles of the input waveform, the configuration is nonlinear with respect to the high-frequency component of the input signal. In other words, the output is not directly proportional to the high-frequency input component. On the other hand, if the modulating information (modulation envelope) is taken as the reference input signal, then the configuration operates in a linear manner. This is just another way of saying that the output amplitude variation is directly proportional to the input amplitude variation of the positive portion of the waveform. Thus, a diode detector operates normally as a linear detector, but insofar as its rectifying action is concerned, it operates as a nonlinear arrangement.

1-9 TRIGGERED-SWEEP OSCILLOSCOPE CONTROLS

A summary of controls and indicators for a typical triggered-sweep oscilloscope is presented in Fig. 1-40. Although two dozen items are noted, only a few controls usually need adjustment when a particular test is made. Readers who are not fully familiar with triggered-sweep oscilloscope operation should give careful attention to these definitions and descriptions. Hands-on experience is almost as important as study of

16. EXT SYNC/HOR jack. Input terminal for external sync or external horizontal input.
17. VAR/HOR GAIN control. Fine sweep time adjustment (horizontal gain adjustment when SWEEP TIME/DIV switch 15 is in EXT position). In the extreme clockwise position (CAL) the sweep time is calibrated.
18. TRIG LEVEL control. Sync level adjustment determines point on waveform slope where sweep starts. In fully counterclockwise (AUTO) position, sweep is automatically synchronized to the average level of the waveform.
19. TRIGGERING SLOPE switch. Selects sync polarity (+), button pushed in, or (−), button out.
20. TRIGGERING SOURCE switch. When the button is pushed in, INT, the waveform being observed is used as the sync trigger. When the button is out, EXT, the signal applied to the EXT SYNC/HOR jack 16 is used as the sync trigger.
21. TVV SYNC switch. When button is pushed in the scope syncs on the vertical component of composite video.
22. TVH SYNC switch. When button is pushed in the scope syncs on the horizontal component of composite video.
23. NOR SYNC switch. When button is pushed in the scope syncs on a portion of the input waveform. Normal mode of operation.
24. FOCUS control. Adjusts sharpness of trace.

8. VARIABLE control. Vertical attenuator adjustment. Fine control of vertical sensitivity. In the extreme clockwise (CAL) position, the vertical attenuator is calibrated.
9. AC vertical input selector switch. When this button is pushed in the dc component of the input signal is eliminated.
10. GND vertical input selector switch. When this button is pushed in the input signal path is opened and the vertical amplifier input is grounded. This provides a zero-signal base line, the position of which can be used as a reference when performing dc measurements.
11. DC vertical input selector switch. When this button is pushed in the ac and dc components of the input signal are applied to vertical amplifier.
12. V INPUT jack. Vertical input.
13. ⏚ terminal. Chassis ground.
14. CAL⊓ jack. Provides calibrated 0.8V p-p square wave output at the line frequency for calibration of the vertical amplifier.
15. SWEEP TIME/DIV switch. Horizontal coarse sweep time selector. Selects calibrated sweep times of 0.5μ SEC/DIV to 0.5 SEC/DIV in 19 steps when VAR/HOR GAIN control 17 is set to CAL. Selects proper sweep time for television composite video waveforms in TVH (television horizontal) and TVV (television vertical) positions. Disables internal sweep generator and displays external horizontal input in EXT position.

1. POWER ON toggle switch. Applies power to oscilloscope.
2. INTENSITY control. Adjusts brightness of trace.
3. Scale. Provides calibration marks for voltage and time measurements.
4. Pilot lamp. Lights when power is applied to oscilloscope.
5. ◆ POSITION control. Rotation adjusts horizontal position of trace. Push-pull switch selects 5X magnification when pulled out, normal when pushed in.
6. ◆ POSITION control. Rotation adjusts vertical position of trace.
7. VOLTS/DIV switch. Vertical attenuator. Coarse adjustment of vertical sensitivity. Vertical sensitivity is calibrated in 11 steps from .01 to 20 volts per division when VARIABLE 8 is set to the CAL position.

Figure 1-40 Summary of controls and indicators for a triggered-sweep oscilloscope. (*Courtesy of* B&K Manufacturing Co., Inc., Division of Dynascan Corp.)

waveform analysis and applications. Therefore, readers are encouraged to obtain a triggered-sweep oscilloscope and to perform various applications described in following chapters. This combination of study and practical experience facilitates learning and assists in retention of the facts.

2

AUDIO-AMPLIFIER MALFUNCTIONS AND WAVEFORM ANALYSIS

2-1 GENERAL CONSIDERATIONS

Although no industry standards have been established, it is generally agreed that high-fidelity reproduction involves uniform frequency response from at least 20 Hz to 20 kHz (within ± 1 dB), with less than 1 percent harmonic distortion. An intermodulation distortion rating of less than 1 percent is also recognized as a high-fidelity requirement. A harmonic-distortion test is made with a sine-wave input, usually at a frequency of 1 kHz. On the other hand, an intermodulation-distortion test is made with a pair of sine-wave input frequencies, typically 60 Hz and 6 kHz. Intermodulation distortion is of basic importance in high-fidelity operation, because all voice and musical tones consist of a fundamental frequency with an array of harmonic frequencies, as exemplified in Fig. 2–1. Specific examples of ōō and ō sound waveforms are shown in Fig. 2–2.

    Careful distinction should be made between harmonic distortion and frequency distortion, or between intermodulation distortion and frequency distortion. In other words, harmonic distortion and intermodulation distortion result from nonlinear amplification. On the other hand, frequency distortion simply denotes lack of uniform frequency response

39

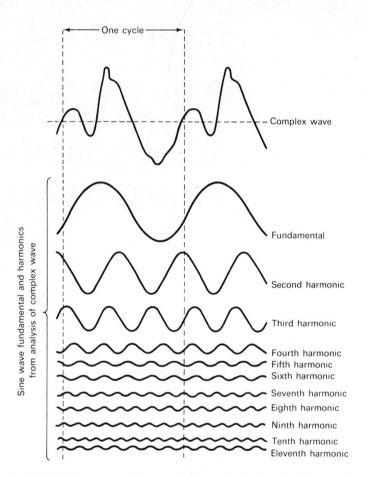

Figure 2–1 Typical tone waveform and its analysis into fundamental and harmonic frequencies.

over a given range. Thus, an amplifier could have negligible harmonic or intermodulation distortion, although its frequency distortion was excessive. A high-fidelity amplifier that has an acceptably low value of harmonic or intermodulation distortion may be operated with various forms of intentional frequency distortion, as shown in Fig. 2–3. As detailed subsequently, a high-fidelity amplifier that has an acceptable frequency response may be intentionally operated with various percentages of harmonic distortion in specialized applications. For example, audio

SECT. 2-1 / GENERAL CONSIDERATIONS

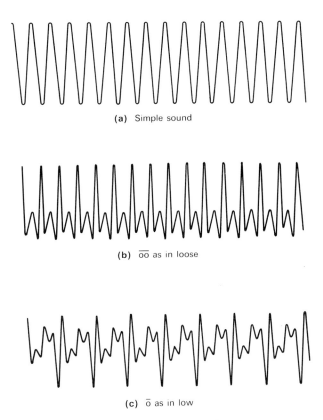

(a) Simple sound

(b) $\overline{oo}$ as in loose

(c) $\bar{o}$ as in low

Figure 2-2 Examples of two vowel-sound waveforms.

expansion and *compression* processes are associated with the introduction of more or less harmonic distortion.

Either an audio oscillator or an audio sweep generator can be used to check the frequency response of an audio amplifier. An advantage of an audio sweep generator (Fig. 2-4) is that the complete frequency response of the amplifier is displayed on the oscilloscope screen, instead of the response's being determined by a series of measurements at various frequencies. It is helpful to utilize a storage-type oscilloscope in this application, inasmuch as a slow horizontal sweep speed should be employed to avoid transient distortion. Note also that the horizontal amplifier in the scope should have DC response. In the example of Fig. 2-4(b), the amplifier exhibits a noticeable amount of low-frequency

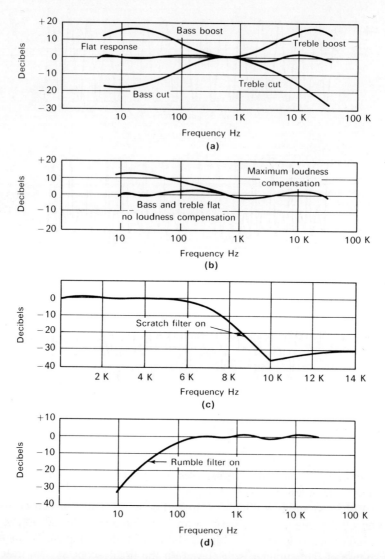

Figure 2–3 Typical amplifier frequency response curves: (a) flat response with effects of bass and treble controls; (b) flat response with effect of loudness control; (c) frequency response of a scratch filter; (d) frequency response of a rumble filter.

boost. Otherwise, its frequency response is reasonably uniform. Observe that the amplifier output terminals must be loaded with a power resistor of suitable resistance value. Power amplifiers are typically rated for 8

SECT. 2-1 / GENERAL CONSIDERATIONS

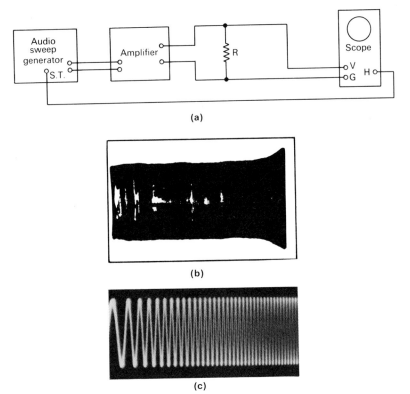

Figure 2-4 Frequency response check for an audio amplifier: (a) test setup using audio sweep generator and oscilloscope; (b) scope display of typical frequency-response pattern; (c) expanded section of audio sweep signal.

ohms output load, whereas preamplifiers are ordinarily rated for 100,000 ohms output load.

A basic frequency response check of an amplifier is made at maximum rated power output. Some amplifiers have poorer frequency response at high power output than at low power output. Hence, it is good practice to measure frequency response at maximum rated rms power output. To calculate the power output value, the peak-to-peak voltage of the sine wave displayed on the scope screen is divided by 2.83 to obtain its equivalent rms voltage value. Then this rms voltage is squared; division of this square value by the load-resistance value gives the rms power value in watts. For example, if the sine-wave peak-to-peak voltage were 8.5 volts, the corresponding rms voltage would be 3 volts, approximately. In turn, the square of this rms voltage is 9. If the load resistance were 4 ohms, the corresponding rms power would be 2.25 watts. Note that the bandwidth of a high-fidelity amplifier is customarily measured within the limits of ±1 dB of the output level at 1 kHz. In addition, the −3 dB bandwidth of the amplifier may be specified. This is the frequency range between the −3 dB points on the frequency response curve, measured at maximum rated rms power output, as shown in Fig. 2–5(a). In other words, the amplifier is driven to maximum rated power output at mid-frequency. Figure 2–6 tabulates basic audio-amplifier tests.

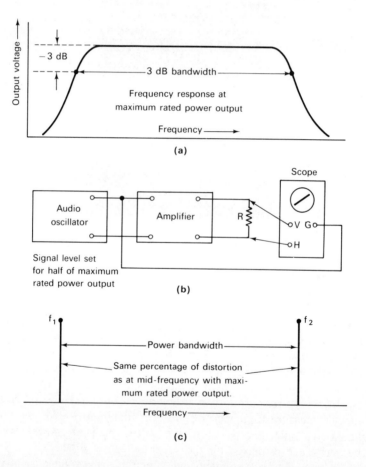

Figure 2–5 Distinction between 3-dB (half-power) bandwith and power bandwidth: (a) 3-dB bandwidth is measured at maximum rated power output; (b) test setup for determining power bandwidth of amplifier; (c) definition of power bandwidth.

SECT. 2-2 / POWER BANDWIDTH MEASUREMENT

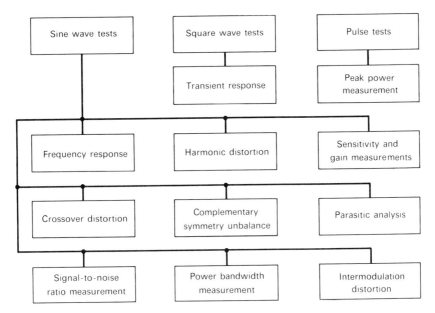

Figure 2-6 Basic audio amplifier tests.

2-2 POWER BANDWIDTH MEASUREMENT

High-fidelity power bandwidth is defined as the frequency range between two frequencies, f_1 and f_2, at which the harmonic distortion at a power level 3 dB below maximum rated power output starts to exceed the mid-frequency distortion at maximum rated power output. A test setup for determination of the power bandwidth of an amplifier is depicted in Fig. 2–5(b), and the significance of f_1 and f_2 is shown in Fig. 2–5(c). It is instructive to observe that small percentages of harmonic distortion are readily apparent in Lissajous patterns, as exemplified in Fig. 2–7. If distortion is negligible, the displayed pattern is a straight diagonal trace. Distortion of 1 percent can be seen in the curvature of the trace, particularly when it is checked against a straightedge. Higher percentages of distortion are evident at a glance. Although it is possible to calculate the percentage of distortion that is displayed in a Lissajous pattern, the measurements and calculations are somewhat tedious. Therefore, Lissajous tests are employed primarily for comparison purposes, as in the measurement of power bandwith. Note that the test setup in Fig. 2–5(b) places stringent demands on the vertical and horizontal amplifiers in the oscilloscope, although the audio oscillator need not be a low-distortion

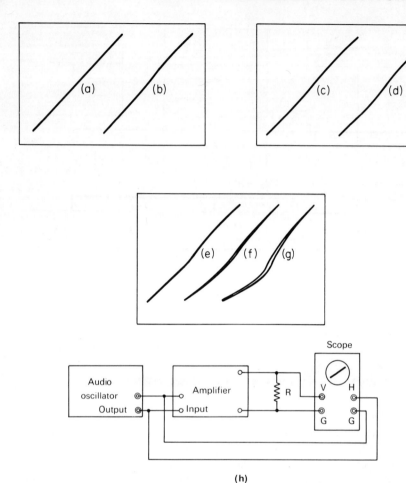

Figure 2–7 Lissajous patterns showing various percentages of harmonic distortion: (a) negligible; (b) one percent; (c) 1.5 percent; (d) two percent; (e) three percent; (f) five percent; (g) ten percent; (h) test setup.

type of instrument. Both the vertical amplifier and the horizontal amplifier in the oscilloscope must have significantly less distortion than the amplifier under test.

Proceed by adjusting the attenuator in the audio oscillator (Fig. 2–5) to provide an amplifier power-output level equal to the amplifier's rated power output. This adjustment may be made at a frequency of 1 kHz.

SECT. 2-3 / MUSIC-POWER WAVEFORMS 47

Then, note the amount of curvature in the displayed Lissajous pattern; for reference, a tracing may be made on the CRT face with a grease pencil. Next, reduce the setting of the attenuator in the audio oscillator to provide an amplifier power-output level 3 dB below the amplifier's rated power output. Then observe the changing curvature in the displayed Lissajous pattern at frequencies of 12, 15, 20, 30, 50, 100, 200, 500, 1000, 2000, 5000, 10,000, 20,000, 30,000, 50,000, 70,000, and 100,000 Hz. Within this range of frequency, two frequencies (f_1 and f_2) will be found that produce the same amount of curvature in the displayed Lissajous pattern as was recorded on the CRT face with a grease pencil. In turn, the frequency range from f_1 to f_2 is equal to the power bandwidth of the amplifier.

2-3 MUSIC-POWER WAVEFORMS

Waveforms such as those illustrated in Figs. 2–1 and 2–2 are voltage waveforms. It is occasionally desired to display corresponding power waveforms. Since the instantaneous power in a waveform is equal to e^2/R, a voltage waveform must be changed into a voltage-squared waveform in order to represent a power waveform. A voltage-squared waveform is produced by multiplying a voltage waveform by itself in an op-amp arrangement, as explained in the previous chapter. In this regard, it is instructive to consider *music-power* waveforms and ratings. Music-power values are based on the peak-power capability of an amplifier. This type of power-value rating is often specified because music waveforms tend to have comparatively high peak values with comparatively low average (rms) values.

With reference to Fig. 2–8, a comparison of voltage and power waveforms for a sine wave and for a musical tone consisting of a fundamental with a substantial third harmonic, note that the ratio of peak power to average or rms power is quite different in the two waveforms. It follows that in practice, the limitation on the power-output capability of a high-fidelity amplifier is the instantaneous peak-power value that it can provide during reproduction of musical passages. Note that the peak-power value of a 10-watt sine wave is 20 watts. On the other hand, a musical tone that has 20 watts of peak power may have less than 5 watts of average (rms) power. Moreover, musical passages generally include many tones that are often played simultaneously. Each of these tones has a certain frequency. On successive peak coincidences, their peak voltages add together, whereas their peak power follows a square law with respect to their sine-wave components. In other words, a

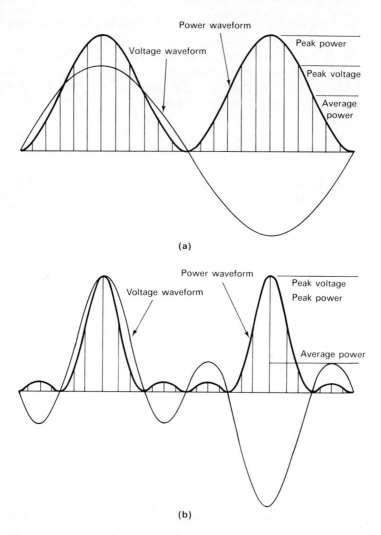

Figure 2–8 A comparison of voltage and power waveforms: (a) voltage, average power, and peak power in a sine wave; (b) voltage, average power, and peak power in a sine wave with a substantial third harmonic.

double peak voltage for any sine-wave component corresponds to a quadrupled peak-power value. Figure 2–9 exemplifies these voltage and power relations. *Music power is defined as the short-term power that is available from an amplifier for the reproduction of program material.* A music-power rating exceeds the corresponding rms power

SECT. 2-3 / MUSIC-POWER WAVEFORMS

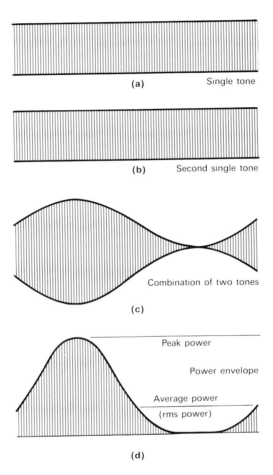

Figure 2-9 Example of peak and average power relations for a combination of two tones with different frequencies.

rating to a greater or lesser extent, depending chiefly upon the regulation of the power supply for the amplifier.

Next, consider three tones, each of which has an average or rms power value of 0.5 watt and a peak-power value of 2 watts. If the amplifier load impedance is 8 ohms, each of the three peaks will have an amplitude of 4 volts. Occasionally, the three peaks will coincide, resulting in a peak value of 12 volts. In turn, the corresponding power value becomes 18 watts, whereas the average or rms power value is 1.5 watts, as exemplified in Fig. 2-10. Furthermore, if an electronic organ tone is under consideration, and a pedal tone is included, it may introduce

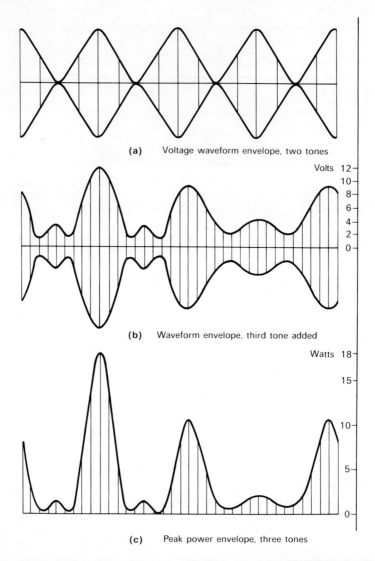

Figure 2–10 Voltage and power waveforms for a combination of three tones.

another 5 watts of average (rms) power with perhaps 18 watts of peak power, also at 12 volts. In turn, a 24-volt peak will result, with a peak-power value of 72 watts and an average (rms) power value of 6.5 watts, as exemplified in Fig. 2–11. In turn, it is evident that the most descriptive power rating for an amplifier that processes musical passages

SECT. 2-4 / EVALUATION OF WAVEFORM DISTORTION

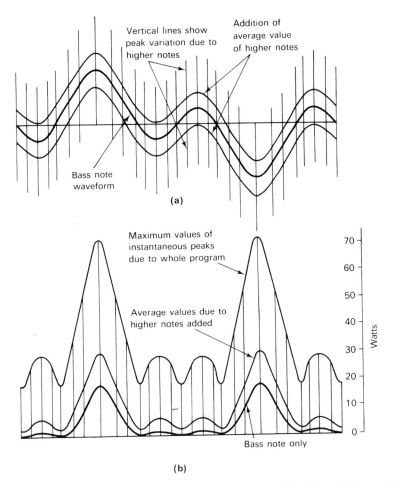

Figure 2–11 Voltage and power waveforms for a combination of three mid-range notes and one bass note.

is a peak-power capability rating. Various high-fidelity amplifiers are rated both for rms power output with a sine-wave input and for music-power output with a pulse input.

2-4 EVALUATION OF WAVEFORM DISTORTION

An amplifier is customarily rated for steady-state distortion, in terms of percentage of harmonic distortion and/or percentage of intermodulation distortion. Amplifiers may also be rated for transient distortion, in terms of ability to reproduce a virtually undistorted square waveform. Other forms of distortion may also be encountered. For example, a high-fidelity amplifier may be rated for its maximum hum-and-noise level output. In other words, the sound output from an amplifier can be distorted by excessive noise and/or hum. A typical noise waveform is illustrated in Fig. 2–12. Hum waveforms generally have a fundamental frequency of either 60 Hz or 120 Hz, and consist of a sine wave plus various harmonics. A typical high-fidelity amplifier is rated for less than 0.5 percent harmonic distortion from 20 Hz to 20,000 Hz, and for at least -60 dB hum-and-noise level with an input signal level of 10 mV rms. This rating denotes that, if a 10-mV signal is applied to the amplifier, an output signal will be obtained that exceeds the hum-and-noise level by at least 1000 times. For example, if the signal voltage across the load is 1 volt and the signal is then switched off, the measured hum-and-noise normally measures less than 0.001 volt.

Although a high-fidelity amplifier normally is free from parasitic oscillation, malfunctioning may develop that results in parasitic oscillation, particularly at maximum rated power output. Parasitic oscillation causes distorted sound output, and sometimes imposes excessive power-dissipation demands on amplifier devices. In turn, overheating and device failure can also result. Parasitic oscillation can be readily observed with an output waveform check, as shown in Fig. 2–13. With the amplifier driven to maximum rated power output, the output waveform is inspected for a "bulge" along its excursion. Although it is possible for parasitic oscillation to occur over any part of the signal cycle, the "bulge" commonly appears at one of the peaks in the waveform.

Next, consider the indication of harmonic distortion in sine-wave patterns, as exemplified in Fig. 2–14. In these examples, the positive peak of the sine waveform is progressively compressed. Note that it is practically impossible to see a distortion of 1 percent in a sine wave.

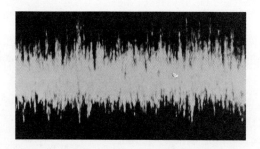

Figure 2–12 A typical noise waveform.

SECT. 2-4 / EVALUATION OF WAVEFORM DISTORTION   53

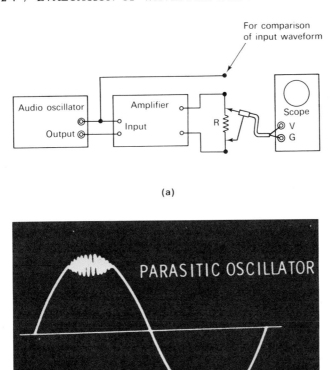

Figure 2–13 Oscilloscope check for parasitic oscillation: (a) test setup; (b) typical display of parasitic oscillator activity.

Comparatively high percentages of distortion, such as 10 or 15 percent, are readily perceptible in a sine-wave display. For this reason, when small percentages of distortion are to be looked for, it is preferable to employ the Lissajous-pattern test setup depicted in Fig. 2–7. If the percentage of harmonic distortion needs to be measured precisely, a harmonic-distortion meter is employed, as shown in Fig. 2–15. It is helpful to connect an oscilloscope at the output of the harmonic-distortion meter as shown in the diagram. An oscilloscope is used to display the waveform of the distortion products when the distortion meter has been adjusted to filter out the fundamental component of the test signal.

Harmonic distortion may consist entirely or chiefly of second-harmonic contamination. Or the distortion may be entirely or chiefly

54 CHAP. 2 / AUDIO-AMPLIFIER MALFUNCTIONS

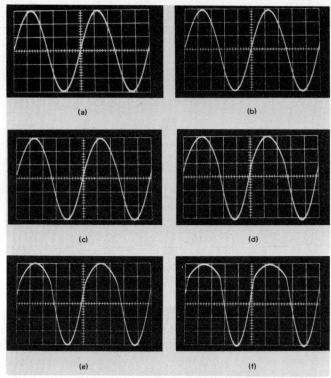

Figure 2-14 Sine wave with various percentages of distortion: (a) 1 percent; (b) 3 percent; (c) 5 percent; (d) 10 percent; (e) 15 percent; (f) 20 percent.

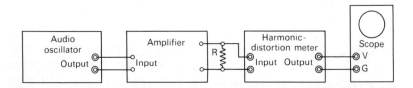

Figure 2-15 Distortion meter and oscilloscope arrangement for checking harmonic distortion.

third-harmonic contamination. Again, a combination of second-harmonic and third-harmonic distortion components may be present. Although higher-order harmonics may be (and often are) present, their amplitudes are ordinarily quite small by comparison. Second-harmonic distortion

SECT. 2-4 / EVALUATION OF WAVEFORM DISTORTION

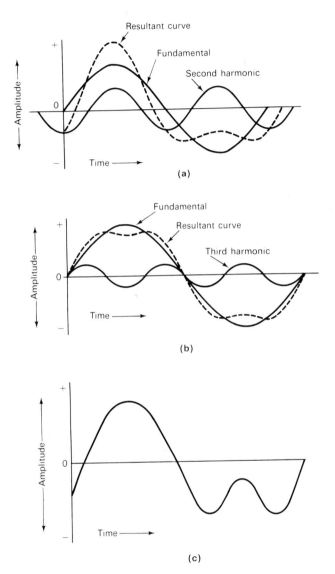

Figure 2-16 Basic second- and third-harmonic distortions: (a) second-harmonic distortion; (b) third-harmonic distortion; (c) combined second- and third-harmonic distortions.

produces a resultant waveform such as that depicted in Fig. 2-16(a). In the arrangement of Fig. 2-15, the harmonic-distortion meter will be tuned to reject the fundamental component, leaving the second-harmonic component to be displayed on the scope screen. Third-harmonic distortion is exemplified in Figure 2–16(b). Distinction is made between second and third harmonics on the basis of their frequencies; in turn, it is advantageous to use an oscilloscope that has triggered sweep with a calibrated time base. If both second-harmonic and third-harmonic distortion are present, the distortion-products waveform has the typical shape depicted in Fig. 2–16(c).

Some forms of distortion increase as the power output of an amplifier is increased. Thus, harmonic distortion is ordinarily greatest when the amplifier is operated at maximum rated power output. On the other hand, crossover distortion, exemplified in Fig. 2–17, increases as the power output of an amplifier is decreased. It is also a form of harmonic distortion, and can be measured with a harmonic-distortion meter. Note that hum-and-noise distortion also increases as the power output of an amplifier is decreased. It can be measured with a harmonic-distortion meter. Since the meter cannot distinguish between various kinds of distortion, troubleshooting procedures are greatly facilitated by employing an oscilloscope at the output of a harmonic-distortion meter.

Transient distortion in high-fidelity amplifiers is usually analyzed by square-wave tests, as shown in Fig. 2–18. A square-wave repetition rate of 2 kHz is typical, with the input signal level set to drive the amplifier to maximum rated power output. For example, suppose that

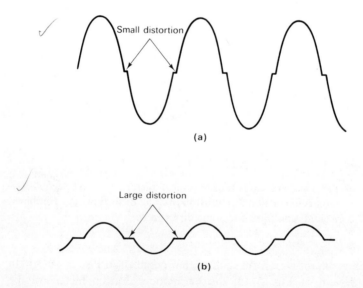

Figure 2–17 Examples of crossover distortion: (a) output at higher amplitude; (b) output at lower amplitude.

SECT. 2-4 / EVALUATION OF WAVEFORM DISTORTION

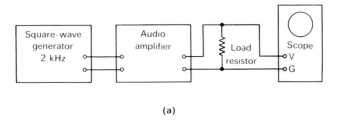

(a)

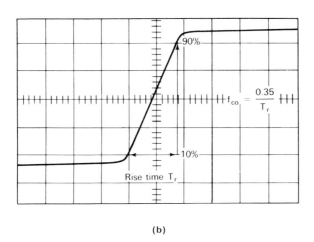

(b)

Figure 2–18 Square-wave test of audio amplifier: (a) test setup; (b) leading-edge display and bandwidth relation.

the amplifier is rated for an rms power output of 25 watts, and that the load resistor has a value of 4 ohms. In turn, maximum rated output will be obtained with a DC voltage of 10 volts across the load resistor. In terms of equivalent square-wave voltage, the input signal level is set to provide a square-wave amplitude of 20 volts peak-to-peak across the load resistor. As exemplified in Fig. 2–18(b), the rise time of a reproduced square wave is related to the high-frequency cutoff as shown in the inset. This measurement requires that the scope have triggered sweep with a calibrated time base.

In many characteristics, a square wave may be considered as the sum of a series of sine waves, as depicted in Fig. 2–19. Sharpness of the corners in a square wave requires reproduction of high-frequency harmonics. On the other hand, if the fundamental frequency is attenuated, the top of the reproduced square wave will sag. Observe that the fundamental and all harmonics depicted in Fig. 2–19 are in phase (go through

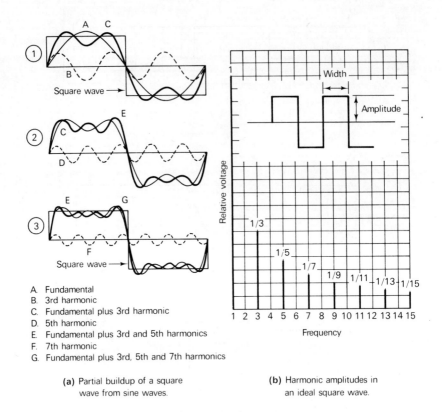

A. Fundamental
B. 3rd harmonic
C. Fundamental plus 3rd harmonic
D. 5th harmonic
E. Fundamental plus 3rd and 5th harmonics
F. 7th harmonic
G. Fundamental plus 3rd, 5th and 7th harmonics

(a) Partial buildup of a square wave from sine waves.

(b) Harmonic amplitudes in an ideal square wave.

Figure 2–19 Synthesis of a square wave from fundamental and harmonic frequencies: (a) partial buildup of a square wave from sine waves; (b) harmonic amplitudes in an ideal square wave.

zero simultaneously). This is a requirement for display of a flat-topped waveform. In other words, if the low-frequency components of the reproduced square wave lead the high-frequency components, the top of the displayed waveform will slope downhill to the right. On the other hand, if the low-frequency components of the reproduced square wave lag the high-frequency components, the top of the displayed waveform will slope uphill to the right. It will also be observed in (1), (2), and (3) of Fig. 2–19 that the rise time of the waveform becomes faster as more harmonics are introduced. Conversely, if the higher harmonics are attenuated in a reproduced square wave, the rise time becomes slower, accordingly.

Next, observe the basic types of square-wave distortion exemplified in Fig. 2–20. It is evident that the shape of a reproduced square wave

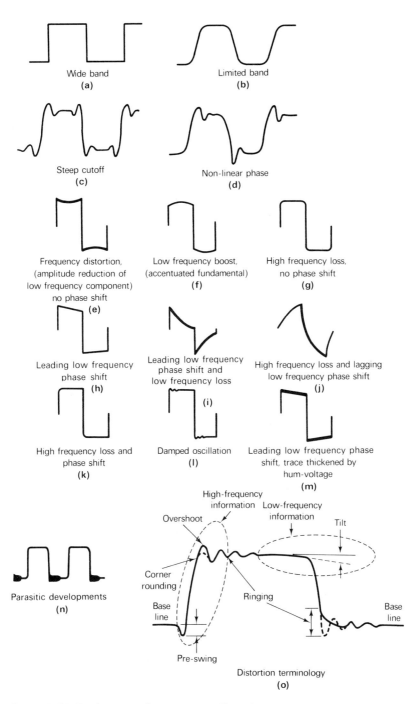

Figure 2–20 Basic types of square-wave distortion.

depends on both the frequency response and the phase characteristic of the amplifier under test. A high-fidelity amplifier normally has a reasonably linear phase characteristic over most of its frequency range. However, all amplifiers exhibit a nonlinear (curved) phase characteristic as the low-frequency cutoff point is approached, and as the high-frequency cutoff point is approached. Whether the reproduced square wave will be noticeably affected by these nonlinear phase regions depends upon whether significant frequency components of the square wave have frequencies that fall in a nonlinear phase region. As an illustration, the reproduced square wave depicted in Fig. 2–20(h) has a top that slopes downhill to the right, because the fundamental frequency of the square wave happens to fall in the low-frequency nonlinear phase interval of the amplifier's frequency response curve. Again, the reproduced square wave depicted in Fig. 2–20(k) has diagonal corner rounding, because significant high-frequency components of the square wave happen to fall in the high-frequency nonlinear phase interval of the amplifier's frequency response curve.

A specified 2-kHz square-wave response for a high-fidelity amplifier is shown in Fig. 2–21. This amplifier has a 1-dB bandwidth of more than 40 kHz, and a 3-dB bandwidth of 86 kHz. In turn, the first 11 harmonics of the square wave are reproduced with no more than 1dB attenuation, and harmonics up to the twenty-second are reproduced with no more than 3 dB attenuation. In turn, the amplifier provides high-quality square-wave reproduction. Note that if an amplifier has a comparatively restricted frequency range, such as 20 Hz to 20 kHz within $\pm$ 1 dB, the 2-kHz square-wave reproduction will be impaired accordingly. In other words, only the first five harmonics will be reproduced with no more than 1 dB attenuation, and approximately ten harmonics will be reproduced with no more than 3 dB attenuation. Since significant high-frequency components will fall in the high-frequency nonlinear phase interval of the amplifier's frequency response curve, diagonal corner rounding will become prominent in this case.

As noted previously, the response of a high-fidelity amplifier to a narrow pulse waveform is related to its music-power (peak-power) rating. A pulse generator is required in most situations, although some square-wave generators have facilities for pulse waveform output. If a triggered-sweep oscilloscope is employed, the reproduced pulse can be expanded for closer inspection. A pulse width of 1 millisecond is suitable, with a repetition rate of 100 pps. To determine the peak-power capability of an amplifier, the amplitude of the applied pulse is progressively increased while the shape of the reproduced pulse waveform is observed.

SECT. 2-4 / EVALUATION OF WAVEFORM DISTORTION

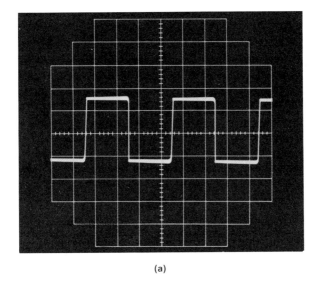

(a)

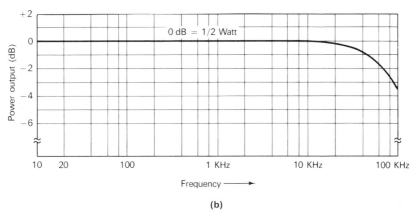

(b)

Figure 2–21 Specified 2-kHz square-wave response for an 8-watt high-fidelity amplifier: (a) reproduced square wave; (b) frequency response of amplifier.

When the peak power capability of the amplifier is passed, the top of the reproduced pulse will lose its rectangular shape, and will become differentiated. To calculate a peak-power value, square the peak voltage of the reproduced pulse and divide the quotient by the ohmic value of the amplifier load resistor.

2-5 AMPLIFIER SENSITIVITY MEASUREMENT AND VOLTAGE GAIN

Amplifier sensitivity is defined as the minimum input voltage that, when applied to the terminals of an amplifier operating under standard test conditions, will produce maximum rated rms power output. A typical test setup is depicted in Fig. 2–13(a). It is customary to utilize a 1-kHz test frequency. First, the generator output amplitude is advanced to obtain maximum rated power output. Then the oscilloscope is transferred to the input terminals of the amplifier, and the ratio of output voltage to input voltage is noted. This ratio is the *voltage gain* of the amplifier, and the rms value of the input voltage is termed the *sensitivity* of the amplifier. Note that some amplifiers have more than one input, as exemplified in Fig. 2–22. In such a case, the input-output measure-

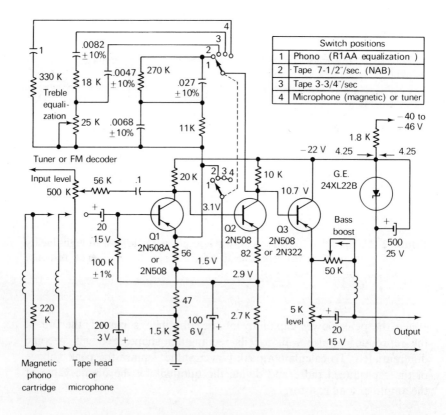

Figure 2–22 Typical amplifier configuration with phono, tape, and tuner inputs. (*Courtesy of* General Electric.)

SECT. 2-7 / INTERMODULATION DISTORTION   63

ment should be repeated for each set of input terminals. Sensitivity and gain values will usually be different for each input.

2-6 UNBALANCED COMPLEMENTARY-SYMMETRY OUTPUT

A basic complementary-symmetry amplifier arrangement is shown in Fig. 2–23. It is in the general category of class-AB amplifiers, except that pnp- and npn-type transistors are utilized in the same stage to provide automatic phase inversion. A complementary-symmetry stage may develop the forms of distortion noted previously, such as crossover distortion. In particular, device or component defects can also cause unbalanced output, wherein the positive peak has a greater amplitude than the negative peak, for example, in the output waveform. Typical waveforms are shown in Fig. 2–24. When unbalanced output is accompanied by peak compression, as in this example, one of the transistors has probably developed collector junction leakage.

2-7 INTERMODULATION DISTORTION

As noted previously, intermodulation distortion is checked with a two-tone signal. A basic test setup with an oscilloscope is shown in Fig. 2–25. Test frequencies of 60 Hz and 6 kHz are often employed, with the

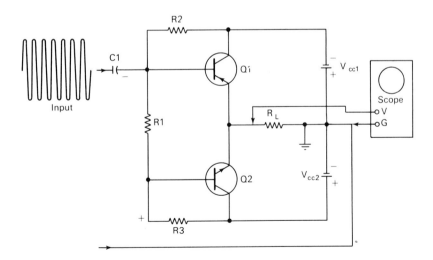

Figure 2–23 Basic complementary-symmetry amplifier arrangement.

64   CHAP. 2 / AUDIO-AMPLIFIER MALFUNCTIONS

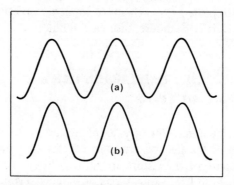

Figure 2-24 Complementary-symmetry amplifier output waveforms: (a) balanced output; (b) unbalanced output.

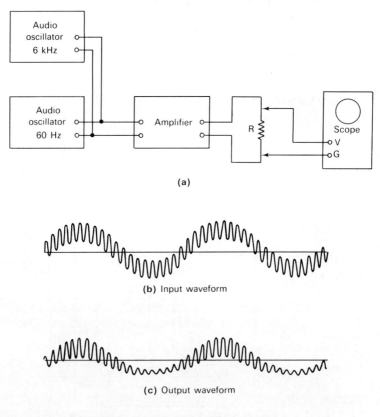

(a)

(b) Input waveform

(c) Output waveform

Figure 2-25 An example of intermodulation distortion.

SECT. 2-7 / INTERMODULATION DISTORTION

6-kHz signal at approximately half the amplitude of the 60-Hz signal. If the amplifier has no intermodulation distortion, the input and output waveforms will be identical. In other words, peak amplitudes will be the same; note that intermodulation distortion causes the 6-kHz component of the output waveform to vary in amplitude over the 60-Hz cycle. This is the key check point to observe. Note that if an amplifier has intermodulation distortion, it will also have harmonic distortion. Both types of distortion tend to increase as the amplifier power output increases. Therefore, it is good practice to make an intermodulation distortion test at the maximum rated rms power output.

Intermodulation distortion can be measured from an oscilloscope display, as shown in Fig. 2–26. A high-pass filter between the amplifier under test and the oscilloscope eliminates the feedthrough 60-Hz component and provides easier measurement of the maximum and minimum values of the AM component. To facilitate measurement of small percentages of IM distortion, the pattern should be displayed at maximum vertical amplitude. On the other hand, the vertical-gain control of the

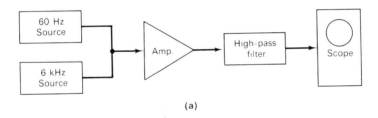

(a)

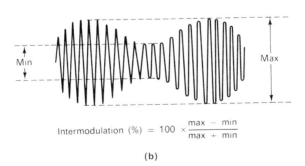

$$\text{Intermodulation (\%)} = 100 \times \frac{\text{max} - \text{min}}{\text{max} + \text{min}}$$

(b)

Figure 2–26 Measurement of percentage of intermodulation distortion: (a) test setup; (b) evaluation of waveform.

oscilloscope should not be advanced to the point that the vertical amplifier becomes nonlinear. For example, in the case of a lab-type oscilloscope that has a graticule 2.5 inches in height, the waveform should not extend vertically beyond the limits of the graticule. Note that some lab-type oscilloscopes have a built-in blanking facility that blanks out any portion of a waveform that extends beyond the limits of precise linearity.

3

STEREO-MULTIPLEX CIRCUIT TESTS

3-1 GENERAL CONSIDERATIONS

Stereo-multiplex circuit operation can be checked to best advantage by waveform analysis. For example, the basic test of stereo receiver or decoder operation concerns the efficiency of left- and right-channel separation. A stereo signal generator is used to energize the unit under test, and the outputs from the L and R channels are checked with an oscilloscope. A vectorscope arrangement may be employed, as depicted in Fig. 3–1. If a stereo receiver is to be tested, a modulated-RF output is applied from the generator. On the other hand, if a stereo decoder

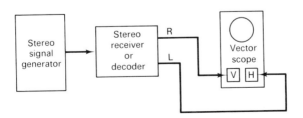

Figure 3–1 Vectorscope method of checking stereo-signal separation.

is to be tested, a composite-audio output is applied. In either case, an R signal is applied without any L signal, and the pattern on the scope screen is observed. Then, an L signal is applied without any R signal, and the resulting pattern is observed. In ideal operation, there would be zero output from the L channel when an R signal is applied; similarly, there would be zero output from the R channel when an L signal is applied. In practice, however, separation is never complete.

Consider the vectorscope stereo-separation patterns shown in Fig. 3–2. If the L channel is energized and separation is complete, a horizontal trace will be displayed on the scope screen. Again, if the R channel is energized and separation is complete, a vertical trace will be displayed on the scope screen. When separation is acceptable, although incomplete, a vertical trace will display a noticeable amount of tilt, as exemplified in Fig. 3–2(c). In case of serious decoder malfunction, with no separation of L and R channels, a diagonal trace will be displayed at an angle of 45 deg with respect to horizontal. Sometimes incomplete separation is accompanied by phase shift. In such a case, a tilted vertical trace will be displayed as an ellipse, as exemplified in Fig. 3–3. If a channel is "dead," there is no corresponding trace on the screen;

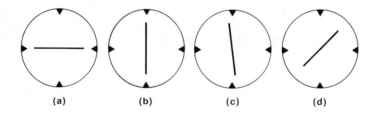

Figure 3–2 Vectorscope stereo-separation patterns: (a) complete separation, L channel energized; (b) complete separation, R channel energized; (c) incomplete separation, L channel energized; (d) zero separation, R channel energized.

Figure 3–3 Incomplete separation with phase shift.

SECT. 3-2 / WAVEFORMS IN STEREO-MULTIPLEX GENERATOR

only a motionless dot is displayed. When channel gain is low, the corresponding trace will lack normal length.

Remember that when a receiver is under test and malfunction is indicated by the scope pattern, either the tuner section or the decoder section may be at fault. Therefore, it is good practice to follow up with a separation test of the decoder with the tuner disconnected. In turn, the test results will show whether the malfunction is in the tuner section or in the decoder section. Figure 3-4 exemplifies typical receiver patterns displayed in normal operation. Instead of utilizing vectorscope indication, some technicians prefer to make a successive pair of tests with an oscilloscope, as depicted in Fig. 3-5. Note that a separation of about 30 dB is considered normal. This is a voltage ratio of approximately 32 to 1. In the example of Fig. 3-5, there is a voltage ratio of 8 to 1 between the R and L channel outputs. In turn, it would be concluded that decoding action is subnormal. A voltage ratio of 8 to 1 corresponds to a separation of 18 dB.

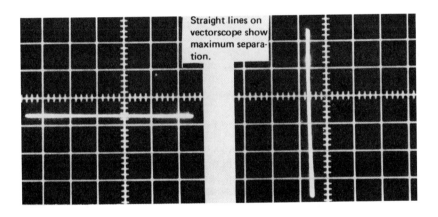

Figure 3-4 Screen photos of typical vectorscope-separation patterns.

3-2 WAVEFORMS IN STEREO-MULTIPLEX GENERATOR

It is instructive to consider the waveforms that occur in stereo-multiplex generator circuitry. With reference to Fig. 3-6, L and R audio composite output signals are provided. These outputs are also available in FM form

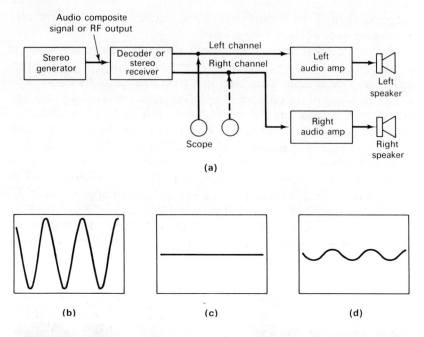

Figure 3–5 Stereo separation test: (a) equipment connections; (b) R channel output; (c) ideal L channel output; (d) L channel output with incomplete separation.

on a 100-MHz carrier. Monophonic and stereophonic signals are provided; the stereo signals can be obtained in their basic form, or mixed with a 19-kHz pilot subcarrier at 5 or 10 percent amplitude. A 67-kHz output signal is also provided for checking the adjustment of SCA traps. An understanding of these signal waveforms follows from a brief consideration of the stereo-multiplex system. A central consideration in system operation is *compatibility,* whereby a monophonic FM receiver processes an FM stereo signal as if it were a monophonic signal. Conversely, compatibility provides an ability for reproduction of monophonic signals by a stereophonic receiver.

Frequency modulation stereo-multiplex transmission and reception provide binaural reproduction in addition to high-fidelity sound. Figure 3–7 shows the basic plan of a stereo-multiplex system. Note that there are two separate audio channels, termed L and R. Since the L microphone picks up a signal that differs from that of the R microphone, different audio waveforms are processed by the L and R channels. FM broadcast channels basically accommodate transmission of monaural signals. To

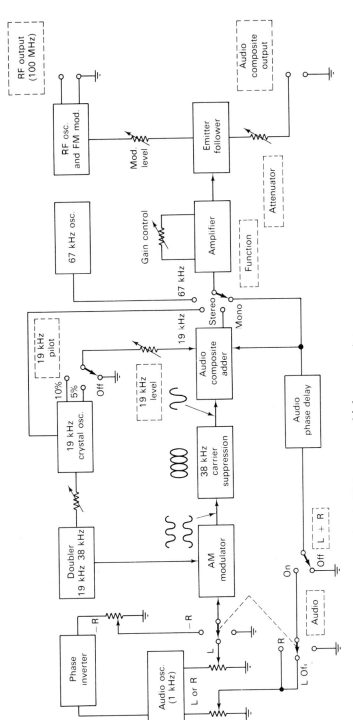

Figure 3–6 Block diagram of a simple FM stereo-multiplex generator.

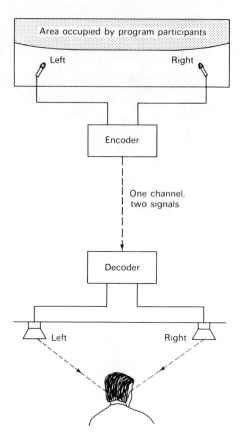

Figure 3–7 Basic plan of a stereo-multiplex system.

transmit a stereo signal in an FM broadcast channel requires a multiplexing technique. Stereo multiplexing permits two separate audio signals to be transmitted in the same FM channel without mutual interference. Thus, FM stereo transmission and reception involve the system depicted in Fig. 3–8, wherein a stereo signal is encoded into the mono signal at the transmitter and is subsequently decoded at the receiver.

In this FM stereo-multiplex signal processing and encoding sequence, the first step consists of the formation of a mono signal from the L and R microphone outputs, as shown in Fig. 3–9. In other words, when the L and R audio signals are added (mixed, or combined), this L + R signal is essentially the same as a conventional mono signal produced

SECT. 3-2 / WAVEFORMS IN STEREO-MULTIPLEX GENERATOR 73

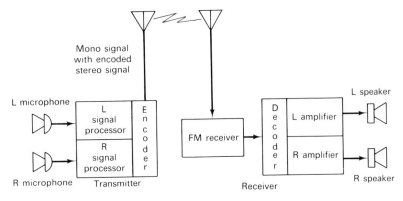

Figure 3–8 FM stereo-multiplex system block diagram.

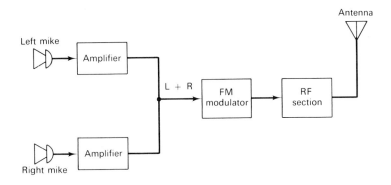

Figure 3–9 Formation of the mono signal.

by a single microphone. In turn, when the L + R signal is applied to the FM modulator, the radiated signal from the antenna is equivalent to a standard mono transmission. It can be received and reproduced by a conventional FM receiver. At this point, one might ask why the mono signal is formed from outputs of two microphones, instead of by a single microphone. The answer is that stereo reproduction requires the availability of an L — R signal, as will be explained.

Figure 3–10 depicts the first step in the encoding process. A 38-kHz CW signal, called the stereo subcarrier, is mixed with the L + R mono signal. This subcarrier does not affect mono reception, inasmuch as its frequency is too high to pass through the audio section of the receiver.

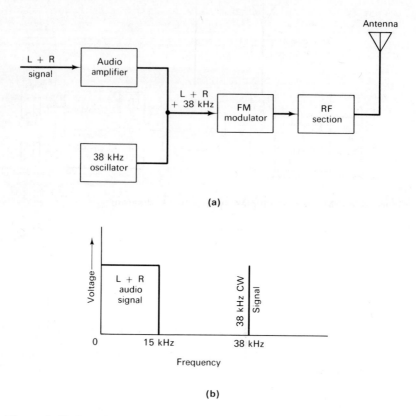

Figure 3–10 Insertion of the stereo subcarrier into the mono signal: (a) basic arrangement; (b) frequency spectrum.

Next, observe that the subcarrier is amplitude-modulated by a second audio signal (A2), as shown in Fig. 3–11. In turn, the subcarrier is accompanied by a pair of sidebands. As before, this modulated subcarrier does not affect mono reception, because all of the sideband frequencies are too high to pass through the audio section of the receiver. However, the L + R signal passes through the audio section and provides normal mono reception. At this point, we have developed a method of encoding a second audio signal into the mono signal. This is the basis for stereo-multiplex transmission and reception, and for stereo-multiplex generator operation (Fig. 3–6).

In addition to the L + R mono signal described above, a stereo-multiplex system also employs an L − R signal. To anticipate subsequent discussions, an L − R signal is required to recover the original L and R

SECT. 3-2 / WAVEFORMS IN STEREO-MULTIPLEX GENERATOR   75

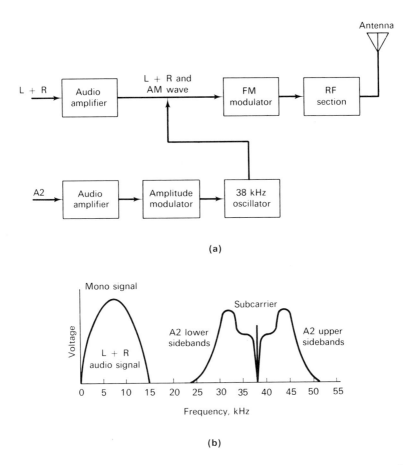

Figure 3–11 Amplitude modulation of the subcarrier by a second audio signal: (a) block diagram; (b) frequency spectrum.

signals at the receiver. In other words, when an L — R signal is added to (mixed with) an L + R signal, a 2L signal is obtained. Again, when an L — R signal is subtracted from (inverted and added with) an L + R signal, a 2R signal is obtained.

Figure 3–12 shows how the L — R signal is formed. Note that the R signal is passed through a polarity inverter, such as a transistor operated in the CE mode. In turn, when the output from the polarity inverter is combined with the L signal, an L — R signal is obtained. Representative L, R, — R, L + R, and L — R waveforms are shown in Fig. 3–13. Frequency modulation stereo-multiplex transmission is then

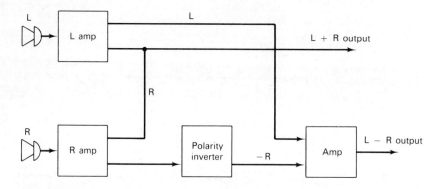

Figure 3–12 Formation of the L − R signal.

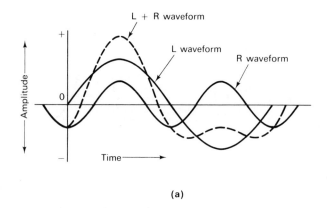

(a)

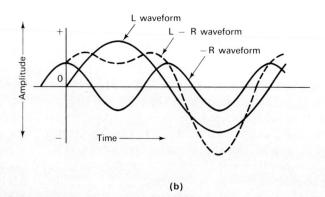

(b)

Figure 3–13 Example of L + R and L − R waveforms: (a) formation of L + R waveform; (b) formation of L − R waveform.

SECT. 3-2 / WAVEFORMS IN STEREO-MULTIPLEX GENERATOR 77

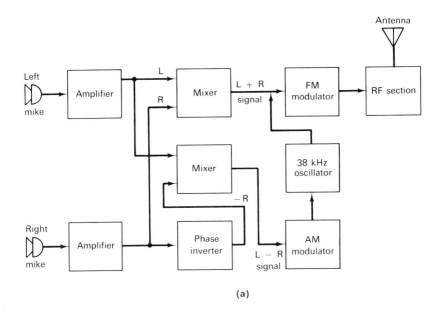

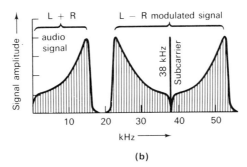

Figure 3–14 FM stereo-multiplex transmitter: (a) block diagram; (b) frequency spectrum.

accomplished with the arrangement shown in Fig. 3–14. This is a completely usable system. However, it has a minor disadvantage in that the 38-kHz subcarrier represents an appreciable fraction of the available power, and thereby detracts from the power that could otherwise go into the L + R signal and the L − R sideband signal. Therefore, in practice, the 38-kHz subcarrier is suppressed at the transmitter, and a low-power pilot subcarrier is inserted, as depicted in Fig. 3–15. Note that the low-power pilot subcarrier operates at 19 kHz, where it is well spaced away from the L + R signal and the L − R sideband signal. To anticipate

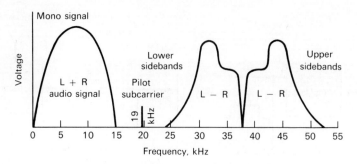

Figure 3–15 The 38-kHz subcarrier is suppressed, and a 19-kHz pilot subcarrier is inserted.

subsequent discussion, the 19-kHz pilot subcarrier is passed through a frequency doubler at the receiver to reconstitute the original 38-kHz subcarrier and stereo signal.

In any FM stereo-multiplex reception system, the 38-kHz subcarrier must be regenerated from the 19-kHz pilot subcarrier. In turn, the 38-kHz subcarrier is inserted into the L — R sideband signal in order to reconstitute the complete L — R signal. Unless the L — R signal is reconstituted, the audio waveform will be seriously distorted. To understand this requirement, it is helpful to observe the waveforms that occur in stereo-multiplex circuitry. Consider the situation in which only the R microphone is energized at the transmitter, so that the L microphone output is zero. Figure 3–16 shows the waveforms that are produced, up to the point that the 38-kHz subcarrier is suppressed. Note that the L — R modulated signal waveform at this point has an envelope corresponding to the L — R signal. If this modulated waveform were transmitted without modification, L — R signal reconstitution would not be required.

On the other hand, as noted previously, the 38-kHz subcarrier is suppressed in the L — R signal prior to transmission. Figure 3–17 shows the result of suppressing the subcarrier. A double-frequency modulation envelope is developed, and this envelope no longer has the shape of a true sine wave. In turn, the L — R signal will be reproduced in seriously distorted form, unless the missing 38-kHz subcarrier is inserted into the L — R sideband signal at the receiver. Observe in Fig. 3–18 that when the R signal is combined with the L — R sidebands, the composite stereo signal is obtained; this is an R composite stereo signal, as provided by a stereo-multiplex generator. Then, when the absent 38-kHz subcarrier is inserted, the reconstituted R signal waveform appears as shown

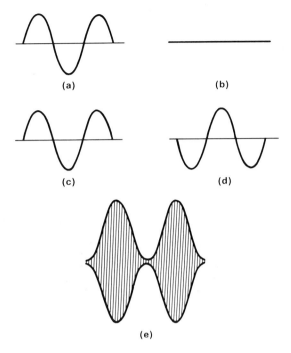

Figure 3–16 Basic waveforms with R microphone only energized: (a) output from R microphone; (b) output from L microphone; (c) L + R signal; (d) L − R signal; (e) L − R signal modulated on 38-kHz subcarrier.

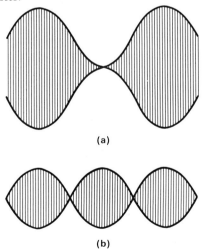

Figure 3–17 Change in shape of amplitude-modulated wave resulting from suppression of carrier: (a) carrier with both sidebands; (b) both sidebands with carrier suppressed.

in Fig. 3–19. When this waveform is passed through an AM detector, the original R audio signal is obtained. This is the basis of the stereo-

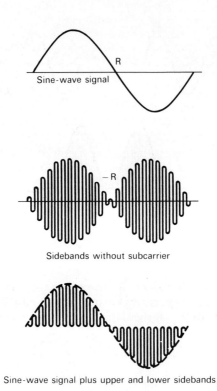

Figure 3–18 Development of R signal waveform.

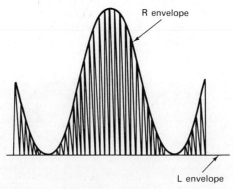

Figure 3–19 Reconstituted R signal waveform.

SECT. 3-2 / WAVEFORMS IN STEREO-MULTIPLEX GENERATOR 81

multiplex decoding process. Note that the composite stereo signal also contains a low-level pilot subcarrier. This pilot subcarrier modifies the waveshape to some extent, as illustrated in Fig. 3–20.

Next, consider the situation in which the R microphone is energized by a certain audio frequency, and the L microphone is energized by a higher audio frequency. If the foregoing waveform processing steps are observed for the additional signal, the reconstituted composite signal is developed as shown in Fig. 3–21. Note that the R envelope waveform is all positive, and that the L envelope waveform is all negative. Advantage is taken of these polarities in one of the basic methods of stereo-multiplex decoding in order to separate the L and R signals. This is accomplished by employing two AM detectors, with their diodes polarized oppositely. In turn, one of the detectors develops the L envelope, and the other detector develops the R envelope. Another method of decoding utilizes a subcarrier reinsertion configuration called a *switching bridge*. This arrangement employs four diodes that operate to develop the L and R audio outputs at the same time that the 38-kHz subcarrier is being reinserted.

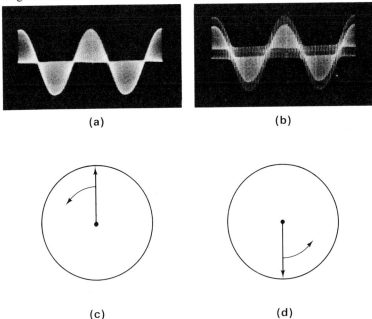

Figure 3–20 Stereo-multiplex test waveforms: (a) R signal without pilot subcarrier; (b) R signal with pilot subcarrier; (c) phase of R signal HF component; (d) phase of L signal HF component.

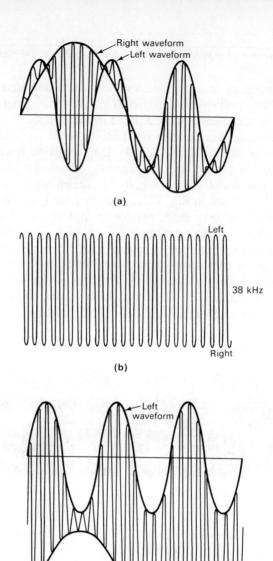

Figure 3–21 Stereo-multiplex subcarrier reinsertion process: (a) incoming L and R signal with suppressed subcarrier; (b) locally generated subcarrier; (c) L and R signal waveform with subcarrier reinserted.

3-3 STEREO-MULTIPLEX WAVEFORM ANALYSIS

Receiver service data generally specify normal peak-to-peak voltages and waveshapes for stereo-multiplex waveforms. Malfunctions result in subnormal waveform amplitude and/or waveshape. Of course, if the local subcarrier oscillator is operating off-frequency, there will also be a frequency error in the distorted waveforms. Normal waveforms depend not only upon the absence of component and device defects, but also upon correct alignment of tuned circuits. As an illustration, IF misalignment can result in phase shift, with the result that the output waveform from the FM detector has baseline curvature, as exemplified in Fig. 3–22. This misalignment condition causes poor separation of the L and R signals. Again, misalignment of the FM detector circuits results in distortion of the audio waveforms, in addition to poor separation. Waveform compression or clipping is a typical symptom in this situation, with a large increase in percentage of harmonic distortion.

In addition to conventional RF and IF tuned circuitry, some stereo-multiplex receivers employ the bandpass-and-matrix system of decoding. This mode of decoding includes a bandpass filter that must be correctly aligned to obtain good separation of the L and R signals. Figure 3–23 shows a standard frequency response curve for the bandpass filter section in a stereo decoder. Alignment adjustments are critical, because the phase characteristic of the filter becomes nonlinear near the ends of the passband. If the filter is misaligned and the bandwidth is subnormal, some

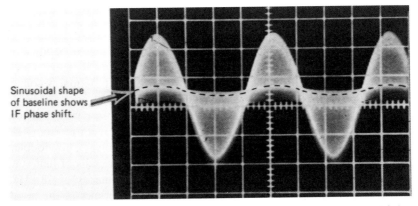

Figure 3–22 IF misalignment evidenced in reproduction of stereo-multiplex test signal. (*Courtesy of* Sencore.)

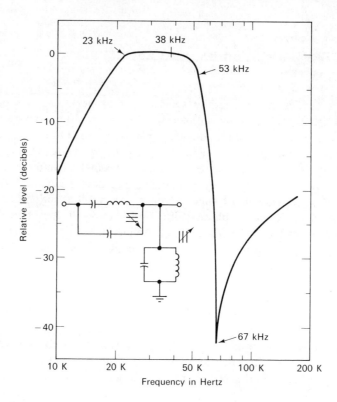

Figure 3–23 Standard frequency response curve for the bandpass filter circuit in a stereo decoder. (*Courtesy of* Heath Co.)

components in the L − R sidebands undergo phase shift. In turn, the subsequent matrixing action becomes impaired; the L − R signal becomes mixed with the L + R signal, and separation becomes unsatisfactory. On the other hand, if the filter is misaligned and the bandwidth is abnormal, higher-frequency components of the L + R signal enter the L − R circuit. Again, the subsequent matrixing action becomes impaired and separation becomes unsatisfactory.

Key troubleshooting waveforms for the subcarrier regenerator section are illustrated in Fig. 3–24. Note that the output waveform from the 19-kHz amplifier is passed through a frequency doubler. This doubler stage operates as a full-wave rectifier, and its output consists of half-sine waves. These half-sine waves have a fundamental frequency of 38 kHz. Tuned circuits in the 38-kHz amplifier reject the harmonics of the input waveform. In turn, the output from the 38-kHz amplifier is a good sine waveform. This CW signal must have normal amplitude; otherwise, the

SECT. 3-4 / NOTES ON NOISE WAVEFORMS

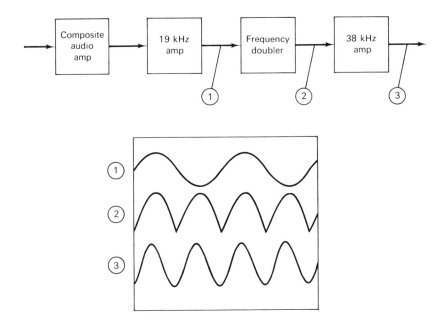

Figure 3-24 Key checkpoints in the subcarrier regenerator section.

L — R signal will be incorrectly reconstituted. In turn, the L and R audio outputs will become distorted and separation will become poor. Observe that the sine-wave output waveform from the 38-kHz amplifier does not need to meet high-fidelity requirements (less than 1 percent harmonic distortion). This requirement is relaxed because the waveform has a basic switching function in the decoder bridge circuit. Apart from adequate amplitude, this waveform needs only to have precisely timed peaks.

3-4 NOTES ON NOISE WAVEFORMS

When an FM receiver is tuned between channels, the noise level is appreciable. On the other hand, when the receiver is tuned to a station with a moderately high signal strength, the noise output is greatly reduced. This large drop in noise output is a result of limiter action. In other words, ordinary noise waveforms are generally amplitude-modulated signals of random characteristics. Limiter action clips the amplitude-modulated noise pulses from an FM waveform, thereby quieting the reception process. However, a limiter is ineffective unless an FM carrier is present. Since random noise waveforms are 100 percent modulated, they will pass through a limiter, regardless of the clipping level. On the other hand, when an FM carrier is tuned in, the AM noise pulses then "ride on" the FM wave envelope. In turn, if the limiter device is biased to clip at a level below the incoming FM wave envelope, virtually all of the noise pulses will be rejected in the limiter stage. When an abnormal noise waveform is displayed at the output of a limiter, with the receiver tuned to a moderately strong FM signal, it is likely that the limiter device is defective, or that an associated component defect has caused incorrect bias.

Random noise waveforms contain all frequencies within the pass band of the circuit under consideration. For example, if an audio amplifier has a pass band from 20 Hz to 20 kHz, random noise waveforms at the output of the amplifier will contain all frequencies within this range. A random noise waveform consists entirely of transient pulses that have varying heights and widths. Accordingly, a noise waveform is continually changing and is never stationary at any time. Noise waveforms produce a rushing sound from a speaker. With all other things being equal, an amplifier that has a pass band of 10 kHz will pass half as much random noise as an amplifier that has a pass band of 20 kHz. Therefore, good noise reduction is a more important consideration in high-fidelity systems than in utility systems that have restricted bandwidth. Note that it is impractical to evaluate the bandwidth of an audio system by analysis of the noise waveform at its output terminals.

Although a random noise waveform contains all frequencies within the pass band of the circuit under consideration, it does not necessarily have the same intensity at all frequencies. If an amplifier or system has a uniform frequency response, the random noise waveform at the output terminals will have the same intensity at low, medium, or high frequencies within the passband. This type of noise waveform is called "white" noise. On the other hand, if the frequency response is nonuniform and a bass boost is introduced, for example, the random noise output will have a comparatively great intensity at low frequencies. This type of noise waveform is called "red" noise. Again, if the frequency response is nonuniform and a treble boost is introduced, for example, the random noise output will have a comparatively great intensity at high frequencies. This type of noise waveform is called "blue" noise. Note that "red," "white," and "blue" noise waveforms have distinctively different sound characteristics. In other words, a "red" noise waveform produces a comparatively low-pitched rushing sound, whereas a "blue" noise waveform produces a comparatively high-pitched rushing sound. A "white" noise waveform produces a medium-pitched rushing sound.

3-5 FM RECEIVER ALIGNMENT

Alignment procedures are employed only after all troubleshooting has been completed. The only exception to this rule is when it is known that the set-owner has tampered with the alignment adjustments. FM receiver sweep-alignment procedure utilizes the test setup depicted in Fig. 3-25. The basic steps that are involved are:

1. Connect an FM sweep generator to the mixer input terminal of the FM receiver. Adjust the generator for a 10.7-MHz center sweep and a deviation of approximately ± 200 kHz.
2. Connect the sweep output terminal of the generator to the external horizontal-input terminal on the oscilloscope, and set the scope controls for external horizontal sweep.

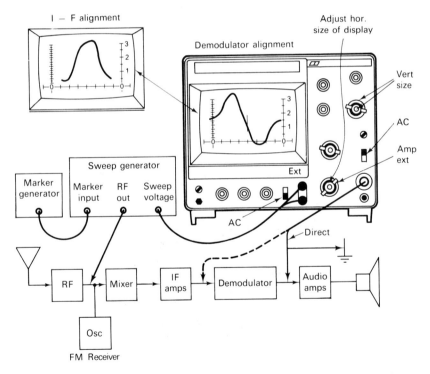

Figure 3-25 FM receiver sweep-alignment setup. (*Courtesy of* B&K Manufacturing Co., Inc., Division of Dynascan Corp.)

3. Connect the vertical-input probe of the oscilloscope to the demodulator test point in the FM receiver. This point is usually located in the last IF stage preceding the discriminator or ratio detector.
4. Adjust the oscilloscope's vertical and horizontal gain controls for a display similar to that shown in Fig. 3–25. Curve evaluation is facilitated by setting the horizontal-gain control to a comparatively high position.
5. Set the marker generator precisely to 10.7 MHz. In normal operation, the marker "pip" appears at top center of the response curve.
6. Align the tuned circuits in the IF strip in accordance with the specifications in the receiver service data.
7. Move the oscilloscope vertical-input probe to the demodulator output terminal. In turn, an S curve should be displayed, and the 10.7-MHz marker "pip" should appear exactly in the center of the S curve. Adjust the tuned circuits in the demodulator stage in accordance with the specifications in the receiver service data, so that the marker moves equal distances from the center point as the marker frequency is increased and decreased equal amounts from the 10.7-MHz center frequency.
8. When the marker is tuned toward the center of the S curve, it may become difficult to see. In such a case, an amplitude-modulated marker signal may be employed. As shown in Fig. 3–26, the "pip" becomes more clearly visible. Also,

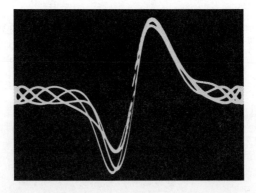

Figure 3–26 Display of amplitude-modulated marker "pip" on FM demodulator S curve.

when the 10.7-MHz marker falls precisely in the center of the S curve, the amplitude of the beat pattern at each end of the baseline is the same.

4

TELEVISION RECEIVER TESTS

4-1 GENERAL CONSIDERATIONS

Although a very wide range of oscilloscope applications and waveform analysis can be utilized in television receiver circuitry, most servicing procedures are limited to the classes of waveforms noted in Fig. 4–1. Technicians recognize two basic groups of waveforms, called *signal waveforms* and *internally generated waveforms*. As an illustration, a video waveform is called a signal waveform, whereas a vertical-sweep waveform is called an internally generated waveform. All waveforms have a voltage component and a current component. For example, a horizontal-sweep waveform has a voltage pattern and a current pattern. It is essential to recognize that a horizontal-sweep voltage waveshape is quite different from a horizontal-sweep current waveshape. In other words, the basic waveform is a power waveform—its voltage and current component waveforms do not necessarily have the same waveshapes, as explained in the previous chapter. Although power waveforms are extensively analyzed in laboratory test procedures, it is customary to restrict analysis to voltage waveforms, with occasional attention to current waveforms in servicing procedures.

A very extensive spectrum of signal waveforms enters the input circuit of the front end (RF tuner). In normal operation, only one of

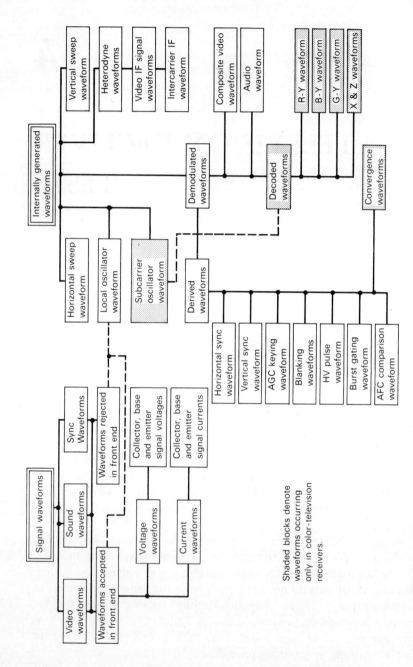

Figure 4-1 Classes of television receiver waveforms.

92

SECT. 4-2 / FRONT-END TESTS

these composite-video modulated-RF waveforms is permitted to pass through to the output of the front end. This is the function of the input-output frequency response of the front end, as exemplified in Fig. 4–2. Observe that the frequency-response curve may be displayed in any one of four aspects. To select a chosen aspect, it is merely necessary to reconnect the pertinent terminals on the CRT terminal board. For example, if terminals 1-2 and 6-7 are cross-connected (Fig. 4–2), the existing display will be inverted. Again, if terminals 4-5 and 9-10 are cross-connected, the existing pattern will be reversed left-to-right. Note in passing that terminals 3 and 8 are in the CRT grid circuit in this example, and are provided for use in external blanking procedures.

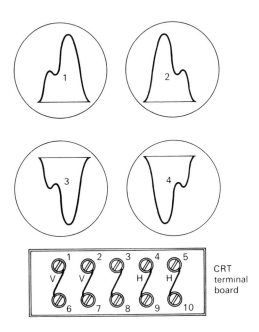

Figure 4–2 Four different aspects of an RF frequency-response curve.

4-2 FRONT-END TESTS

In normal operation, a front end has a specified minimum gain, such as 20 dB in high-band operation, and 26 dB in low-band operation. A front end also has a specified limit on waveshape variation, as exemplified in Fig. 4–3. Since service-type RF sweep generators are not provided with calibrated output attenuators, technicians customarily estimate tuner

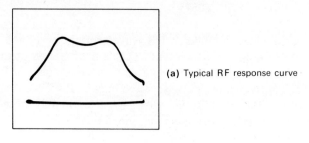

(a) Typical RF response curve

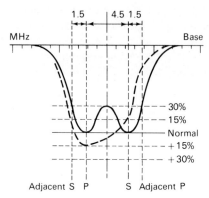

(b) Response-curve limits specified for a television receiver.

Channel number	Picture carrier frequency (MHz)	Sound carrier frequency (MHz)	Receiver VHF oscillator frequency (MHz)
2	55.25	59.75	101
3	61.25	65.75	107
4	67.25	71.75	113
5	77.25	81.75	123
6	83.25	87.75	129
7	175.25	179.75	221
8	181.25	185.75	227
9	187.25	191.75	233
10	193.25	197.75	239
11	199.25	203.75	245
12	205.25	209.75	251
13	211.25	215.75	257

(c) Frequency reference chart

Figure 4–3 Typical RF frequency-response curve, and example of specified limits with operating frequencies: (a) typical R-F response curve; (b) response-curve limits specified for a television receiver; (c) frequency reference chart.

SECT. 4-2 / FRONT-END TESTS

gain in terms of instrument control settings. In other words, after sufficient experience has been acquired in the operation of a particular RF sweep generator and oscilloscope, the technician recognizes "normal" attenuator settings. In turn, if increased generator output and/or increased oscilloscope sensitivity is required in a particular situation in order to obtain reference pattern amplitude on the oscilloscope screen, the technician recognizes that the front end has subnormal gain and in turn requires troubleshooting.

As a practical note, keep in mind that apparent subnormal gain in a front end can also be caused by instrument malfunctions. Thus, if the RF sweep generator has subnormal output, or if the oscilloscope has subnormal vertical-amplifier gain, the front end under test will appear to have subnormal gain. Also, the output amplitude from the generator over the swept band should be uniform (flat) in order to avoid a false appearance of curve distortion. In other words, if a generator malfunction causes the RF output to vary from 100 percent to 20 percent over the swept band, the displayed curve will appear distorted, as exemplified in Fig. 4–4. A practical test for uniformity of generator output is shown in Fig. 4–5. As a general rule, it is desirable that departure from "flatness" be no more than ±1 dB over the swept band.

Receiver service data may specify frequency response in terms of decibels instead of percentage of full voltage at various points on a response curve. To convert amplitude percentages to corresponding decibel values at key points in a pattern, refer to the relations depicted in Fig. 4–6. Note in passing that not all service-type RF sweep generators have sufficient deviation (sweep width) to clearly show the ends of an RF response curve. However, nearly all generators provide return-trace blanking in the form of a zero-volt base line, as exemplified in the upper

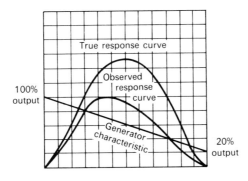

Figure 4–4 Example of distorted response curve owing to nonuniform generator output.

96   CHAP. 4 / TELEVISION RECEIVER TESTS

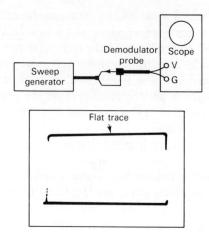

Figure 4–5 Check of generator output uniformity and example of satisfactorily flat trace.

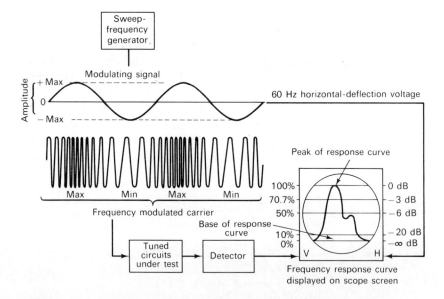

Figure 4–6 Amplitude-decibel relationship for a frequency-response curve.

pattern of Fig. 4–3. In turn, although the ends of the displayed curve are "chopped off," the zero-volt level is indicated in the pattern, enabling a ready estimate of dB (or percentage amplitude) points.

Frequencies along response curves are indicated by means of

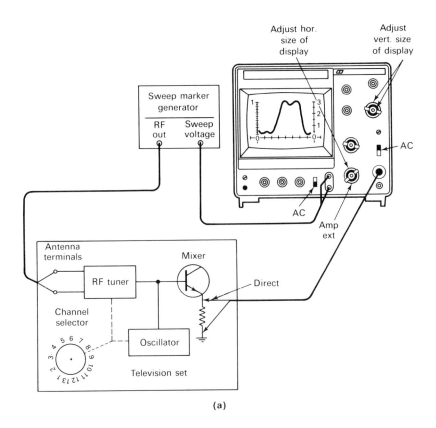

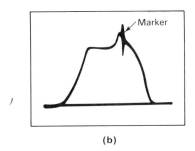

Figure 4–7 Example of a front-end alignment arrangement: (a) test setup; (b) typical curve display with beat marker. (*Courtesy of* B&K Manufacturing Co., Inc., Division of Dynascan Corp.)

markers. Beat markers are often employed, as exemplified in Fig. 4–7. In most cases, the sweep generator contains a built-in marker generator.

However, a separate marker generator is sometimes utilized. In such a case, the output from the marker generator is loosely coupled to the input terminals of the front end. Thus, the ends of the marker-generator output cable may be placed in the vicinity of the input terminals to the front end. There has been a considerable trend toward the use of post-marker injection. Basically, the arrangement depicted in Fig. 4–8 is em-

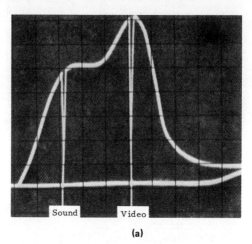

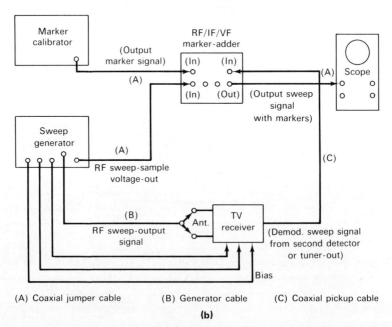

Figure 4–8 Post-injection marking of RF frequency-response curve: (a) example of large post-injection markers; (b) typical test setup.

SECT. 4-3 / SYNC-PULSE WAVEFORM ANALYSIS 99

ployed. A marker waveshaper, often called a marker-adder unit, is utilized to form a beat marker in a separate process, so that the marker signal does not pass through the tuned circuits under test. In turn, the marker signal is injected directly into the oscilloscope input terminals. A considerable advantage is provided by post-marker injection, in that very large markers can be used, if desired, without overloading the circuits under test. Moreover, post-injection markers are always displayed at the same amplitude, and do not become "choked out" at low-amplitude points on a response curve. Figure 4-8 exemplifies large post-injection markers on a response curve.

4-3 SYNC-PULSE WAVEFORM ANALYSIS

Horizontal and vertical sync pulses are informative concerning signal-circuit action. As an illustration, waveform proportions indicate whether the IF amplifier is operating in a linear or a nonlinear manner. Relative amplitudes of the horizontal and vertical sync pulses indicate whether the IF amplifier may have subnormal low-frequency, or subnormal high-frequency response. Bandwidth is indicated by the relative squareness of corners and the degree of rounding in the reproduction of the sync tips. Sloping tops in displayed sync pulses indicate nonlinear phase shift. Although these distortions can occur in front-end processing of the composite video signal, they are introduced by IF-amplifier malfunction in most cases. In other words, the IF amplifier has less bandwidth than the front end, and its frequency-response curve is somewhat critical by comparison.

An idealized IF waveform corresponding to the composite video signal is pictured in Fig. 4-9. Although this waveform has a lower carrier frequency in the IF amplifier than in the front end, its frequency remains too high to be processed by service-type oscilloscopes. Thus, carrier frequencies in the front end range from 55.25 to 211.25 MHz in VHF reception, and the IF picture-carrier frequency is always 45.75 MHz in normal operation. Only lab-type oscilloscopes can process a 45.75-MHz signal. In turn, the waveform depicted in Fig. 4-9 is never displayed as such in service shops. When a technician traces the signal through an IF amplifier, he utilizes a demodulator probe. This type of probe operates as a traveling detector, and changes the IF input signal into a video-frequency output signal. Since nearly all service-type oscilloscopes have a 4- or 5-MHz vertical-amplifier bandwidth, this test method provides a practical expedient. It is, at best, an expedient, inasmuch as conventional demodulator probes impose substantial circuit loading and have limited demodulating capability. For this reason,

100   CHAP. 4 / TELEVISION RECEIVER TESTS

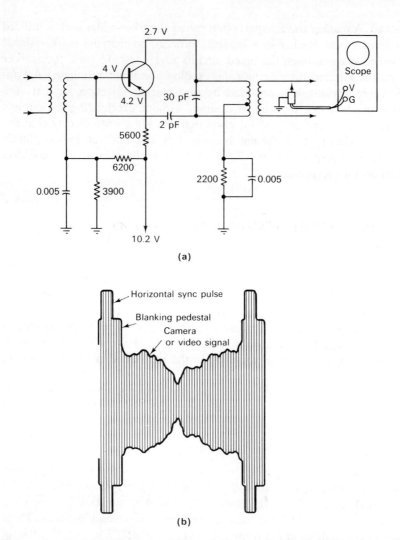

Figure 4–9 IF waveform in normal receiver operation: (a) test setup; (b) normal display.

technicians start IF waveform analysis by evaluating the video-detector output waveform.

First, the video-detector output waveform is checked for amplitude. Typical specified values for various receivers range from 1 to 3 volts peak-to-peak. Note that although the amplitude of the output waveform will change to some extent if the input test-signal level is changed, this

SECT. 4-3 / SYNC-PULSE WAVEFORM ANALYSIS

variation is masked to a considerable extent by AGC action. A typical IF strip has a maximum gain (including the mixer stage) of approximately 80 dB. Thus, the IF section normally provides the larger portion of the signal-channel gain, in addition to the basic selectivity of the picture channel. A tolerance of ±20 percent on sync-pulse amplitude is regarded as normal in usual servicing procedures. Figure 4–10 shows the meaning of this tolerance specification in terms of the sync-pulse display. Departures out of these tolerance limits are regarded as trouble symptoms, requiring follow-up circuit tests to determine the cause of malfunction. Note that the three displays in Fig. 4–10 have the same signal proportions—the sync tip is the same fraction of the pedestal height in each case. Only the total amplitude of the waveform differs from one display to the next.

Next, observe the normal horizontal sync-pulse display illustrated in Fig. 4–11. This video signal represents a test pattern; it has reference-white highlights. In turn, the sync-tip amplitude is approximately one-fourth of the total waveform amplitude. This is the standard proportion of waveform components for a video signal with reference-white highlights. However, this proportion will be maintained through the picture-signal channel only if amplification is linear. For example, if the bias on the last IF-amplifier transistor becomes incorrect, the stage may overload when driven to normal output level. In the case of an IF amplifier, overload always results in sync-tip compression. Overload is most likely to occur in the last IF stage, because the signal level is highest in this stage. However, sync-tip compression may occur in an earlier IF stage, in the event of a serious nonlinearity malfunction. The reason that nonlinear IF amplification always results in sync-tip compression is seen from the waveform shown in Fig. 4–9. In other words, both the positive peak and the negative peak of the waveform consist of sync tips. In turn,

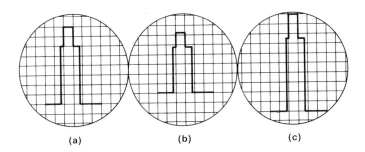

Figure 4–10 Example of ±20 percent tolerance: (a) specified amplitude; (b) −20 percent amplitude; (c) +20 percent amplitude.

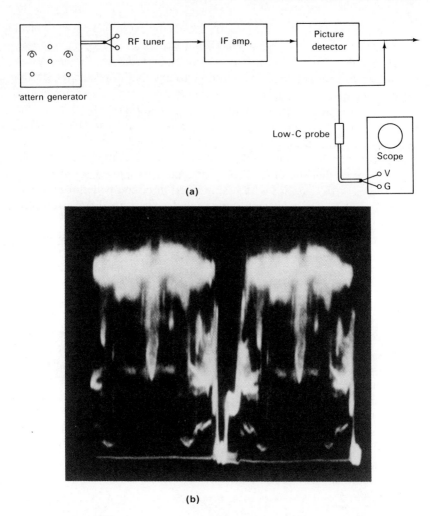

Figure 4–11 Horizontal sync-pulse display: (a) test setup; (b) normal pattern.

whether amplitude nonlinearity involves the positive excursion or the negative excursion of the waveform, the result is to change the normal proportions of the sync tips. Thus, the transistor in the malfunctioning stage may be driven into saturation or into cutoff—in either case, it is the sync-tip component of the waveform that is most affected.

As a practical example of sync-pulse compression or clipping, refer to Fig. 4–12. There is no sharp dividing line between compression and clipping distortion. However, clipping denotes substantially complete removal of the peak component in a waveform, whereas compression

SECT. 4-3 / SYNC-PULSE WAVEFORM ANALYSIS

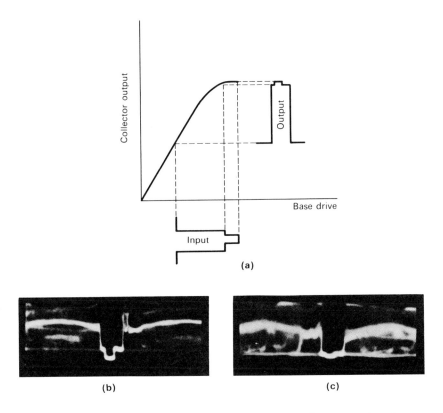

Figure 4–12 Example of sync-pulse clipping: (a) nonlinear transfer characteristic; (b) normal sync-pulse waveform; (c) distorted waveform with sync tip highly compressed.

denotes reproduction of the peak component on an attenuated scale. Note that when a transistor is driven into saturation, the resulting output waveform displays compression distortion. On the other hand, if the transistor is driven beyond cutoff, the resulting output waveform displays clipping distortion. Observe that the sync-pulse waveform illustrated in Fig. 4–12(b) is within normal tolerance, whereas the distorted waveform pictured in Fig. 4–12(c) exhibits a highly compressed sync tip. This distortion involves a serious operating condition, because horizontal sync is lost as a result. Therefore, the troubleshooter must proceed to check amplifier operation in detail, and to correct the component or device defect that is causing sync-tip compression.

Next, it is helpful to note the effect of limited high-frequency response on sync-pulse reproduction. With reference to Fig. 4–13, a horizontal sync pulse will be reproduced with comparatively square corners and fast rise by a picture channel with 4-MHz bandwidth. On the other hand, limited high-frequency response results in corner rounding and slowed rise of the reproduced sync pulse. In turn, this waveform distortion is accompanied by loss of picture detail, although there is no adverse effect on horizontal-locking action. As a practical consideration, note that an IF amplifier cannot reproduce a normal sync-pulse waveform unless it is driven by a normal sync-pulse waveform. Because TV-station video waveforms occasionally depart substantially from the ideal, it is good practice to employ an adequate test-pattern generator in advanced servicing procedures. In turn, distortions in reproduced waveforms will clearly indicate circuit malfunctions.

In addition to details of horizontal sync-pulse reproduction, picture-channel bandwidth is also indicated by relative amplitudes of the horizontal sync pulses and the vertical sync pulses. This type of waveform analysis is exemplified in Fig. 4–14. Horizontal sync pulses and equalizing pulses contain higher-frequency information than do vertical sync pulses. Accordingly, an amplifier must have normal low-frequency response to reproduce the vertical sync pulses correctly. If the amplifier

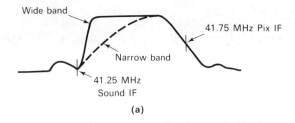

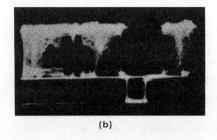
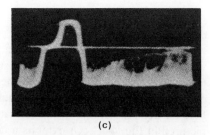

Figure 4–13 Sync-pulse waveshapes versus picture-channel bandwidth: (a) wide-band and narrow-band frequency response; (b) wide-band sync-pulse reproduction; (c) narrow-band sync-pulse reproduction.

SECT. 4-3 / SYNC-PULSE WAVEFORM ANALYSIS

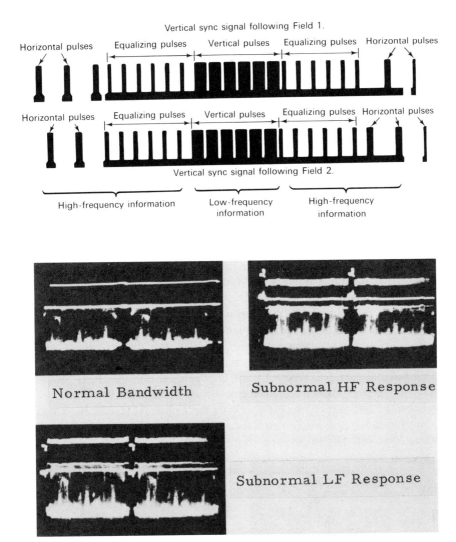

Figure 4–14 High-frequency and low-frequency information in the sync train.

lacks adequate low-frequency response, the vertical sync pulse will become attenuated. On the other hand, an amplifier must have normal high-frequency response to reproduce horizontal sync pulses correctly. If the amplifier lacks normal high-frequency response, the horizontal sync pulse will become attenuated. It follows that the three key waveforms shown in Fig. 4–14 are basic indicators of picture-channel frequency response.

Many comparatively puzzling TV trouble symptoms are caused by entry of hum voltage into various signal and control circuits. This abnormality is readily apparent in a reproduced composite video signal, as exemplified in Fig. 4–15. Although hum usually appears as a linearly mixed waveform component, the technician will also encounter hum-modulated waveforms on occasion. In other words, a composite video signal may be amplitude-modulated by 60-Hz or by 120-Hz hum. This distortion occurs if the supply voltage to the local oscillator is contaminated with excessive ripple voltage, for example. Hum-modulated video waveforms are also produced by excessive hum voltage on an AGC line. Note in passing that a "floating" high-impedance lead will pick up substantial hum voltage. For example, if a gate lead to a field-effect transistor is open-circuited, it is likely to pick up stray-field hum voltage, which is amplified in turn and appears in the drain circuit.

In addition to hum voltage, waveforms may occasionally become contaminated with television interference (TVI). As an illustration, Fig. 4–16 shows an RF/IF frequency-response curve that is mixed with a vertical sync-pulse residue. Note that the picture-carrier and sound-carrier markers on the response curve are produced by the TV station-signal carriers. When TVI is encountered during RF/IF alignment procedures, the channel-selector switch of the receiver should be switched to a vacant channel. Otherwise, the alignment test setup should be moved into a screened room. Note that TVI can also be encountered when front-end alignment procedures are in progress. However, the

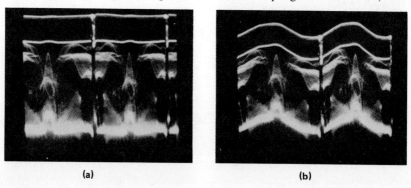

Figure 4–15 Appearance of 60-Hz hum in the video signal waveform: (a) normal display of composite video signal; (b) waveform mixed with 60-Hz hum voltage.

SECT. 4-4 / IF ALIGNMENT PROCEDURE

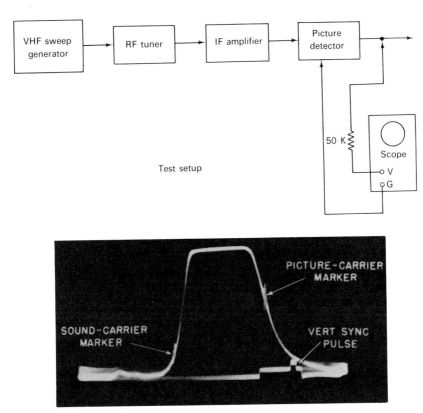

Figure 4-16 TV station interference marks RF/IF response curve.

interference is ordinarily less evident in this case, because the system gain is comparatively low, and a higher-level sweep signal is employed.

4-4 IF ALIGNMENT PROCEDURE

When IF alignment is required, it is important to consult the receiver service data for consecutive steps and detailed instructions. However, the following general procedure illustrates the basic factors that are involved. With reference to Fig. 4-17, connect the output of the sweep generator to the signal-injection point in the mixer stage. Then adjust the sweep generator to sweep the IF frequency band. Note that if the tuner has normal response, the IF strip may be aligned by applying an RF sweep signal to the antenna-input terminals of the receiver, and the

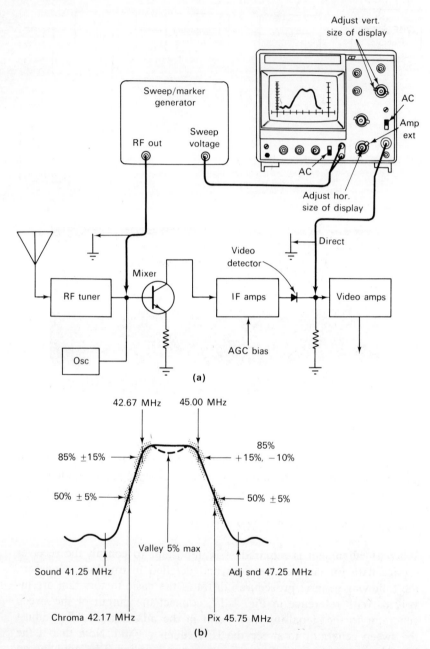

Figure 4–17 Example of IF alignment procedure: (a) test setup; (b) IF response curve, with permissible tolerances noted. (*Courtesy of* B&K Manufacturing Co., Inc., Division of Dynascan Corp.)

SECT. 4-5 / DEFLECTION CURRENT WAVEFORM CHECKS 109

end result will be practically the same. Next, synchronize the oscilloscope with the sweep generator. Connect the ground clip from the oscilloscope's vertical probe to the chassis of the receiver; connect the vertical-input probe of the oscilloscope to the picture-detector output terminal, using a direct cable with a 10-k series "isolating" resistor. This resistor provides RC low-pass action and sharpens the beat-marker indication considerably.

Next, the vertical-gain controls of the oscilloscope are adjusted for approximately two-thirds of full-screen deflection, while the sweep-generator output level is set to a comparatively low value to ensure that the IF amplifier is not overloaded at the peak of the response curve. Overloading results in artificial flat-topping and distortion of the pattern. Then the marker frequencies are set as required to check the key points on the response curve. Note that the AGC bias indicated in Fig. 4–17 is clamped to the value recommended in the receiver service data. Either batteries or a bias box may be employed. Some sweep generators provide an override bias output voltage. Unless the AGC line is clamped, the displayed response curve is likely to be significantly distorted, owing to the tendency of the AGC system to maintain a uniform signal amplitude at the output of the picture detector.

Next, consider sync-pulse reproduction through the video-amplifier section, as depicted in Fig. 4–18. Video-amplifier gain is measured by comparing the input/output signal-amplitude ratio at points (1) and (2). A typical gain specification for a video amplifier is 28 dB. This corresponds to an amplitude ratio of approximately 30 times. Although the input horizontal-sync pulse to the video amplifier is within normal tolerances, the output horizontal-sync pulse occasionally displays excessive distortion owing to sync-pulse compression or to "white saturation," as exemplified in Fig. 4–18. Note that sync-tip compression or clipping and "white saturation" caused by limiting are basically the same processes, except that the former occurs on the opposite polarity peak with respect to the latter. In either case, the most likely defect is incorrect bias on a video-amplifier transistor. When amplifier linearity needs to be checked precisely, a staircase signal is employed, as shown in Fig. 4–19. This test signal is extensively used in laboratory procedures, and is occasionally utilized in service shops.

4-5 DEFLECTION CURRENT WAVEFORM CHECKS

Beam deflection in a picture tube is accomplished by flow of sawtooth current waveforms in the horizontal and vertical deflection coils.

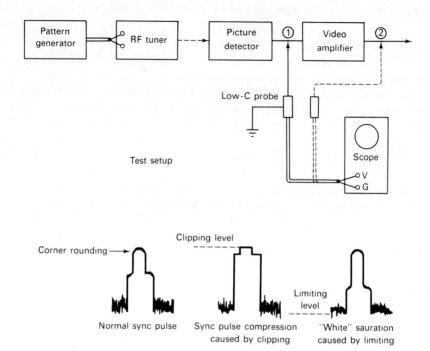

Figure 4–18 Two basic types of pulse distortion that may occur in a video amplifier.

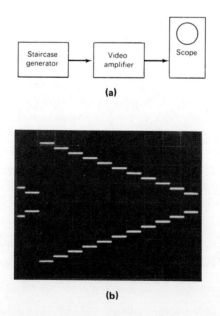

Figure 4–19 Staircase signal checks video-amplifier linearity: (a) test arrangement; (b) typical staircase signal waveform.

SECT. 4-5 / DEFLECTION CURRENT WAVEFORM CHECKS

Scanning linearity depends upon sawtooth waveform linearity. Thus, if the sawtooth-current waveform is not straight, but curved, nonlinear scanning action results. As an illustration, if the ramp portion of the current sawtooth departs by 16 percent from a straight line, the scanning nonlinearity is 16 percent, as depicted in Fig. 4–20. In other words, the percentage of scanning nonlinearity is measured in terms of the peak-to-peak current by which the deflection waveform fails to equal the height of the ideal deflection waveform. Sawtooth current waveforms are displayed to best advantage by means of a current probe, as shown in Fig. 4–21. This is essentially a miniature current transformer with a split core that can be clamped around the pertinent wire lead. In turn, the lead serves as a primary, and the secondary drives the oscilloscope input circuit. Not shown in Fig. 4–21 is a small compensating transistor amplifier that is utilized between the probe and the scope. This amplifier provides uniform frequency response (from 60 Hz to 4 MHz in this example), and also increased sensitivity. A sensitivity of 1 mV per mA is provided by the exemplified probe.

With reference to Fig. 4–22, a sawtooth current flow through resistance produces a sawtooth voltage drop. On the other hand, a sawtooth current flow through pure inductance produces a rectangular or pulse voltage drop. Again, a sawtooth current flow through resistance and inductance connected in series produces a peaked-sawtooth voltage drop. Regardless of the load parameters, scanning linearity requires a linear sawtooth current flow through the load. Horizontal-deflection coils have comparatively small resistance, and can be regarded as an ideal inductive load in the first analysis. On the other hand, vertical-deflection coils have substantial winding resistance, and must be regarded as a series RL load in the first analysis. In addition, both horizontal- and vertical-deflection coils contain appreciable distributed capacitance. Consequently, a deflection coil is a resonant circuit in practice. The horizontal-deflection system normally resonates at approximately 70 kHz. Although distributed capacitance does not change the basic requirement for linear sawtooth current flow through the deflection coils,

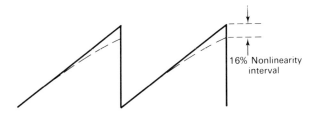

Figure 4–20 Example of 16 percent scanning nonlinearity.

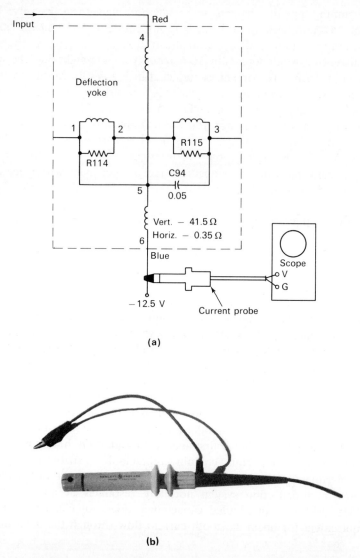

Figure 4-21 Checkout of deflection-current waveform: (a) test setup; (b) appearance of current probe. (*Courtesy of* Hewlett-Packard.)

this parameter does impose adequate damping requirements on the scanning system. Unless the deflection coils are adequately damped, "ringing" distortion will occur in the sawtooth current waveform, as exemplified in Fig. 4-23.

SECT. 4-6 / CONVENTIONAL OSCILLOSCOPE PROBES

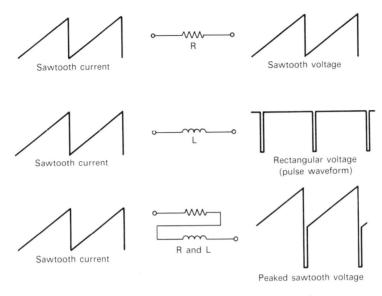

Figure 4-22 Voltage waveforms associated with sawtooth current waveforms in resistive, pure inductive, and resistive-inductive loads.

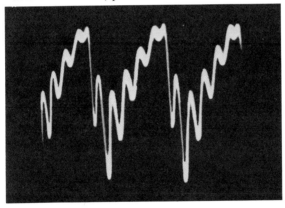

Figure 4-23 Example of severe ringing in a current-sawtooth waveform.

4-6 NOTES ON CONVENTIONAL OSCILLOSCOPE PROBES

Demodulator probes used in IF-amplifier signal-tracing procedures were noted previously. Although various configurations are utilized, the circuit shown in Fig. 4-24 is typical. Note that the demodulating capability of this probe extends to 5 kHz. In turn, only the vertical-sync

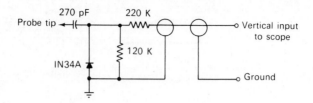

Frequency response characteristics:
 RF carrier range .... 500 kHz to 250 MHz
 Modulated-signal range............ 30 to 5000 Hertz
Input capacitance (approx.) .... 2.25 pF
Equivalent input resistance (approx.):
 At 500 kHz .... 25,000 Ohms
 1 MHz .... 23,000 Ohms
 5 MHz .... 21,000 Ohms
 10 MHz .... 18,000 Ohms
 50 MHz .... 10,000 Ohms
 100 MHz .... 5000 Ohms
 150 MHz .... 4500 Ohms
 200 MHz .... 2500 Ohms
Maximum input:
 AC voltage.......................... 28 Peak volts

(a)

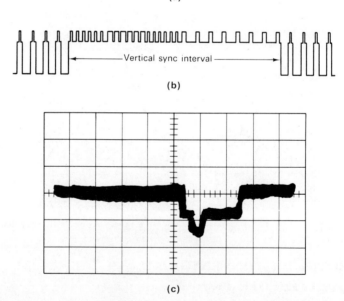

Figure 4-24 Demodulator probe configuration and characteristics: (a) circuit; (b) detail of the vertical-sync interval; (c) probe response to the vertical-sync interval.

SECT. 4-6 / CONVENTIONAL OSCILLOSCOPE PROBES

interval of the composite video waveform produces appreciable output from a demodulator probe. In other words, the fundamental frequency of the horizontal-sync pulse is 15,750 Hz. Because the exemplified probe has an input resistance of approximately 10,000 ohms and an input capacitance of 2.25 pF at IF frequencies, circuit loading is substantial, and the output signal from the probe is not necessarily proportional to the amplitude of the input signal. Moreover, the input capacitance of the probe tends to detune the IF circuit under test, with the result that the apparent amplitude of the output waveform may be either greater or less than is the case in normal operation. Occasionally, the stage under test will be detuned with respect to an adjacent stage in such manner that the stage breaks into oscillation. In such a case, the signal is stopped, and the stage appears to be "dead."

An oscilloscope can be applied satisfactorily with open test leads, or with a simple coaxial cable, in low-impedance low-frequency circuits. For example, no probe is required when waveforms in power-supply circuits are checked. On the other hand, a low-capacitance probe is essential when other than low-impedance circuits are checked. A low-capacitance probe is designed to increase the effective input impedance of an oscilloscope (with its coaxial input cable) by a factor of ten. Thus, if an oscilloscope has a basic input impedance represented by 90 pF shunted by 1 megohm, a standard low-capacitance probe will change these values effectively to 9 pF shunted by 10 megohms. Of course, a tradeoff is involved wherein the output signal amplitude from the low-capacitance probe is reduced to 0.1 of the input signal amplitude. A low-capacitance probe provides an important advantage of greatly reduced circuit loading and minimizes the possibility of spurious waveform distortion from this cause.

A helpful summary of the key test points in a television receiver, with notations concerning the use of a low-capacitance or demodulator probe and the appropriate oscilloscope deflection rate, is presented in Fig. 4–25. As explained previously, direct test leads can be used instead of a low-capacitance probe, provided that the circuit under test has low impedance. A sensitivity increase of ten times is provided by the use of direct test leads.

A low-capacitance probe must be properly compensated to avoid waveform distortion. Compensation is accomplished by adjusting the trimmer capacitor provided inside of the probe. To check the adjustment of the probe, a square-wave test may be employed, as depicted in Fig. 4–26. If the trimmer capacitor is correctly adjusted, a 1-kHz square wave will be reproduced with square corners and a flat top. On the other

116 CHAP. 4 / TELEVISION RECEIVER TESTS

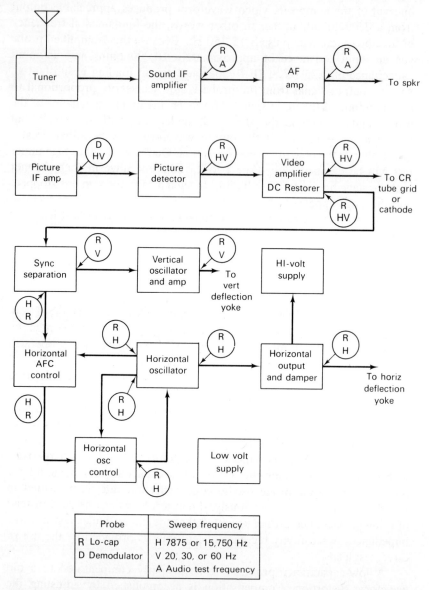

Figure 4-25 Summary of key test points and scope application in a TV receiver. (*Courtesy of* Heath Co.)

hand, if the probe trimmer capacitance is excessive, the reproduced square wave will display overshoot on leading and trailing edges. Again, if the probe trimmer capacitance is insufficient, the reproduced square

SECT. 4-7 / BASIC WAVESHAPING PROCESSES

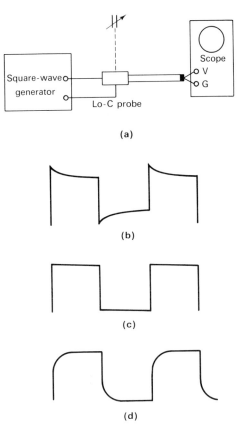

Figure 4-26 Low-capacitance probe adjustment with 1-kHz square wave: (a) test setup; (b) excessive trimmer capacitance in probe; (c) correct trimmer capacitance; (d) insufficient trimmer capacitance.

wave will display diagonal corner rounding. Note that the test frequency is not critical; any square-wave repetition rate between 1 kHz and 10 kHz may be utilized. Of course, the square-wave generator must supply a good square waveshape. Otherwise, the deficiencies of the generator will be falsely charged to the low-capacitance probe.

4-7 BASIC WAVESHAPING PROCESSES

All complex waveforms contain high-frequency and low-frequency components. A practical example of this composition is seen in Fig. 4–27.

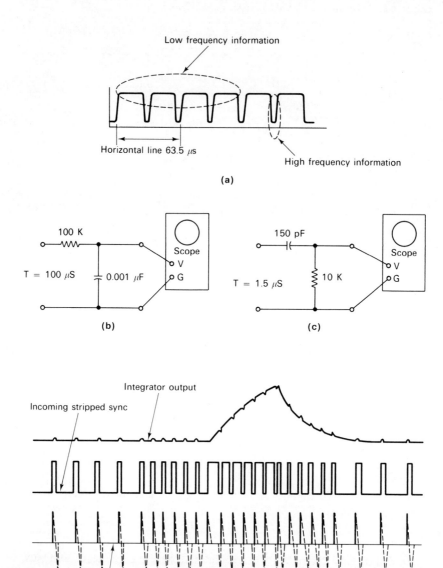

Figure 4–27 Low-frequency and high-frequency information in the sync train: (a) regions of low- and high-frequency information in the vertical-sync pulse; (b) differentiating circuit separates high-frequency component; (c) integrating circuit separates low-frequency component; (d) waveforms at input and at outputs of differentiating and integrating circuits.

SECT. 4-7 / BASIC WAVESHAPING PROCESSES   119

Thus, the general form of the vertical-sync pulse defines its low-frequency information, whereas the serrations in the vertical-sync pulse define its high-frequency information. Horizontal sync pulses carry high-frequency information, and very little low-frequency information. An RC differentiator is a simple form of high-pass filter; in turn, it develops the high-frequency content of the stripped-sync waveform. On the other hand, an RC integrator is a simple form of low-pass filter; accordingly, it develops the low-frequency content of the stripped-sync waveform. As pictured in Fig. 4–27, the low-frequency content of the sync train is displayed basically in the form of an integrated square wave that has a comparatively slow repetition rate. On the other hand, the high-frequency content of the sync train is displayed as a series of narrow spike-shaped pulses. Observe that integration develops the comparatively sustained portion of the input waveform, whereas differentiation develops the rapidly changing portion of the input waveform.

Next, observe the blanking-pulse waveshaping circuit depicted in Fig. 4–28. This circuit changes a peaked-sawtooth waveform into a pulse

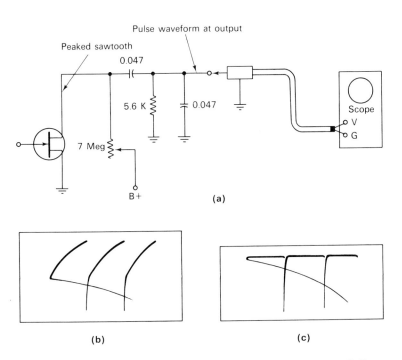

Figure 4–28 Development of a blanking pulse: (a) high-pass RC waveshaping circuit; (b) peaked-sawtooth input waveform; (c) pulse output waveform.

waveform. Note that the high-frequency information in a peaked-sawtooth waveform is contained in its peaking pulse. On the other hand, the low-frequency information in a peaked-sawtooth waveform is contained in its ramp portion. Accordingly, when the peak sawtooth is applied to a high-pass RC waveshaping circuit, the output waveform is a series of pulses. Note that development of a properly shaped output waveform depends upon correct component values in the waveshaping network.

Another RC waveshaping circuit employed in some receiver systems is exemplified in Fig. 4–29. This is a sawtooth peaking arrangement that changes a sawtooth waveform into a peaked-sawtooth waveform. Peaking is accomplished by the series RC branch circuit; the component values are chosen for optimum peaking action. Operation of the peaking circuit is based on the gradual charging of C during the ramp interval. At the start of retrace, the capacitor has acquired a positive charge. As the retrace interval of the output waveform drops rapidly to zero, capacitor C requires a short time to draw electrons through R to bring its terminal voltage to zero. This surge of electron flow through R causes the output lead to develop a negative-going pulse. In turn, the output waveform becomes a peaked sawtooth.

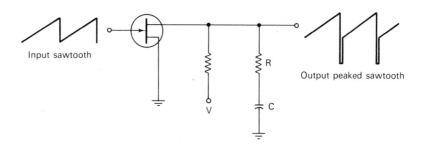

Figure 4–29 RC sawtooth peaking configuration.

4-8 INTERCARRIER-SOUND SECTION WAVEFORMS

In normal operation, the waveform in the 4.5-MHz intercarrier-sound IF section is a frequency-modulated signal that is passed through a limiter stage to remove any amplitude-modulation interference. This FM waveform is then demodulated by a discriminator, ratio detector, or other equivalent arrangement. Various receivers employ some form of

SECT. 4-8 / INTERCARRIER-SOUND SECTION WAVEFORMS

slope detection, such as the delta sound system. This mode of FM demodulation is less familiar to most technicians than the older discriminator and ratio-detector configurations. With reference to Fig. 4–30, the frequency-modulated signal is applied to a resonant load in which the 4.5-MHz point falls halfway down the side of its frequency-response curve. In turn, the load output rises and falls in amplitude as the FM signal increases and decreases in frequency. Accordingly, the output from the slope detector is an FM carrier with an AM envelope. This is the first step in slope detection. The second step consists of passing the output waveform from the tuned load through a conventional AM diode-detector circuit. Thereby, the audio information is demodulated from the AM envelope of the FM carrier wave.

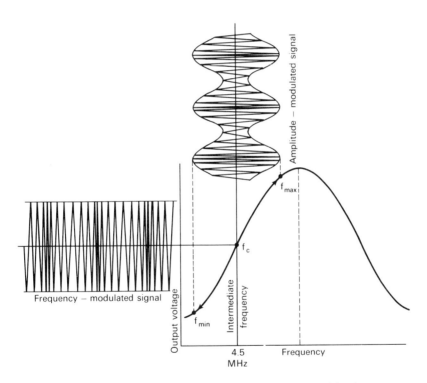

Note: The amplitude – modulated signal is finally demodulated by a diode detector.

Figure 4–30 Waveforms in slope-detection process.

4-9 THREE FORMS OF SYNC-BUZZ INTERFERENCE

Sync-buzz interference in the sound output occurs as a harsh 60-Hz rasping tone. Its intensity may be steady, or it may vary with the background brightness of the displayed picture. In most cases, the amplitude of the sync-buzz tone increases as the volume control is advanced. These characteristics are useful in preliminary analysis of this trouble symptom. Three types of sync-buzz waveforms will be encountered, as pictured in Fig. 4–31. Thus, the sync pulse may appear as a downward modulation of the 4.5-MHz sound signal. Again, the sync pulse may appear as an upward modulation of the sound signal. Or the sync pulse may appear linearly mixed with the sound signal. As before, these characteristics are useful in preliminary analysis when one is troubleshooting a sync-buzz trouble symptom.

Sweep-buzz interference should be clearly distinguished from sync-buzz interference. Although the tone of sweep buzz is very similar to the tone of sync buzz, the source of the interference is different. In other words, sync buzz arises in the picture-signal channel—in the IF amplifier or in the video amplifier. On the other hand, the source of sweep buzz is in the vertical-sweep section. If a decoupling capacitor between the intercarrier-sound section (or the audio section) and the vertical-sweep section becomes open-circuited, for example, 60-Hz vertical-sweep pulses may gain entry into the sound section. Note that when sync buzz occurs, the interference waveform is identifiable as a sync pulse (Fig. 4–31). On the other hand, when sweep buzz occurs, an oscilloscope check will show a simple spike-shaped interference waveform.

4-10 ANTENNA AND TV RECEIVER IMPEDANCE MATCHING TESTS

A check of the impedance match of a lead-in to a TV receiver or to an antenna can be made with a VHF sweep generator, as shown in Fig. 4–32. A double-ended demodulator probe should be utilized, as diagrammed. Note that the length of the lead-in that is used in the test is not critical, although it should be at least one-quarter wavelength at the highest frequency of test. If the lead-in is many wavelengths in extent, the test results will be essentially the same as if it is comparatively short.

SECT. 4-10 / ANTENNA AND TV IMPEDANCE MATCHING TESTS 123

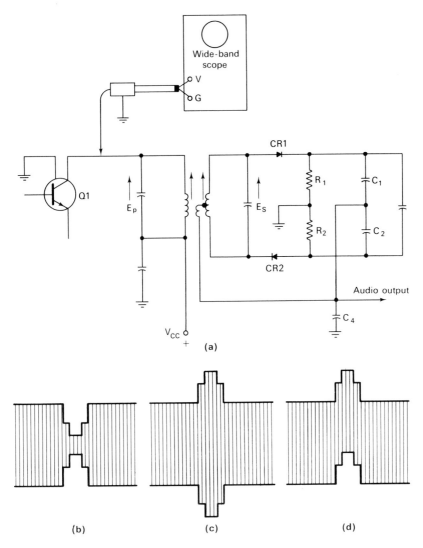

Figure 4–31 Three types of sync-buzz waveforms: (a) test setup; (b) downward modulation; (c) upward modulation; (d) linear mix.

The sweep generator should be tuned to sweep the channel of interest; in most cases, the technician wishes to check the impedance match over all of the VHF channels from 2 through 13. A horizontal trace indicates a good impedance match. On the other hand, dips and humps in the trace indicate an impedance mismatch.

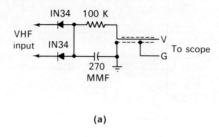

(a)

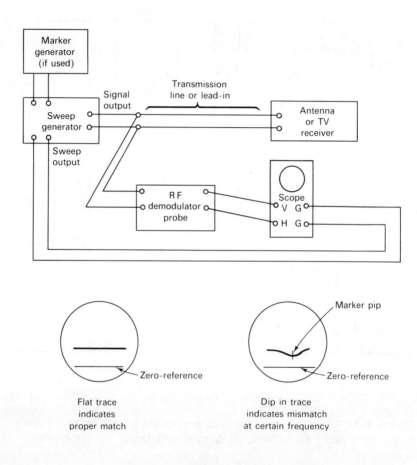

(b)

Figure 4-32 Check of high-frequency impedance match: (a) double-ended demodulator probe; (b) arrangement for checking match of antenna or receiver to lead-in.

SECT. 4-10 / ANTENNA AND TV IMPEDANCE MATCHING TESTS

To measure the absolute value of the mismatch, various composition resistors may be connected to the end of the lead-in instead of the antenna or the TV receiver. When the same magnitude of dip(s) and/or hump(s) is determined by experiment with different values of resistors, the absolute value of the mismatch has been determined. Observe in Fig. 4–32(b) that the double-ended demodulator probe is connected at the end of the line near the sweep generator. If desired, the probe may be connected at the generator terminals. Note in passing that the sweep generator should be of a type that provides a good sine waveform (does not have appreciable harmonic output). Also, the output voltage should be uniform within ± 1 dB over the swept channel. Otherwise, the deficiencies of the generator will be falsely attributed to the receiver or the antenna under test.

5

COLOR-TV RECEIVER TESTS

5-1 GENERAL CONSIDERATIONS

A block diagram of a typical color-TV receiver, showing signal paths, is depicted in Fig. 5–1. Note that the tuner frequency response is essentially the same as in a black-and-white receiver. However, the IF amplifier is aligned for a somewhat greater 6-dB bandwith—3.58 MHz versus 3.50 MHz, for example. Observe that the first video amplifier is followed by a delay line and two more video-amplifier stages. Video amplifiers following the delay line are often called Y amplifiers. A 3.58-MHz color-subcarrier trap is included in the Y-amplifier channel. With reference to Fig. 5–2, the frequency response of a video and/or Y amplifier is checked with a video sweep generator, demodulator probe, and oscilloscope. As is plainly visible in Fig. 5–2(b), the 3.58-MHz subcarrier trap normally provides substantial attenuation of the chroma signal for approximately 0.4 MHz on either side of the subcarrier. Observe in Fig. 5–2(c) that noise interference may be a problem during the frequency-response check, unless the IF amplifier is biased off during the test procedure.

Subcarrier trap action affects the video waveform as exemplified in

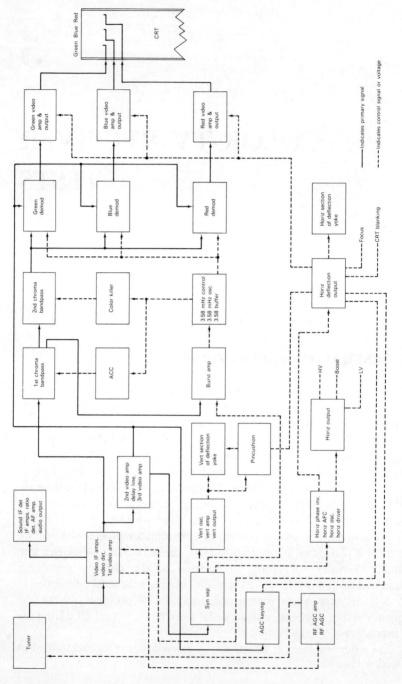

Figure 5-1 Block diagram of a typical color-TV receiver, showing signal paths.

SECT. 5-1 / GENERAL CONSIDERATIONS 129

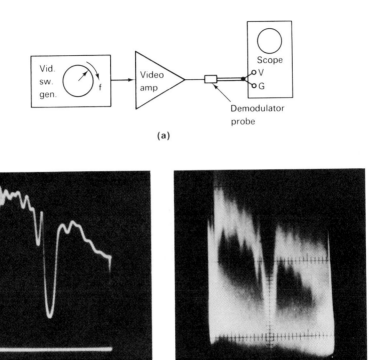

Figure 5-2 Check of Y-amplifier frequency response: (a) test setup; (b) normal display; (c) noise interference.

Fig. 5-3. Normally, the color burst has the same amplitude as the sync tip. If the trap attenuates the chroma subcarrier to two-thirds of its original value (attenuates the subcarrier approximately 3.8 dB), the color burst appears as shown in Fig. 5-3(b). Next, if the trap attenuates the subcarrier to one-half of its original value (an attenuation of 6 dB), the color burst appears as shown in Fig. 5-3(c). Again, if the trap attenuates the subcarrier to one-sixth of its original value (an attenuation of approximately 24 dB), the color burst appears as shown in Fig. 5-3(d). Note in passing that the color subcarrier is attenuated by 6 dB in the IF amplifier of most color receivers. In other words, the subcarrier falls 6 dB down on the IF frequency-response curve, on the opposite side from the picture carrier. In turn, the subcarrier is attenuated 6 dB, compared with its amplitude as provided by old-model receivers in which the subcarrier is placed on top of the IF response curve. Thus, if the subcarrier trap reduces the color burst by 24 dB, the IF amplifier almost

130   CHAP. 5 / COLOR-TV RECEIVER TESTS

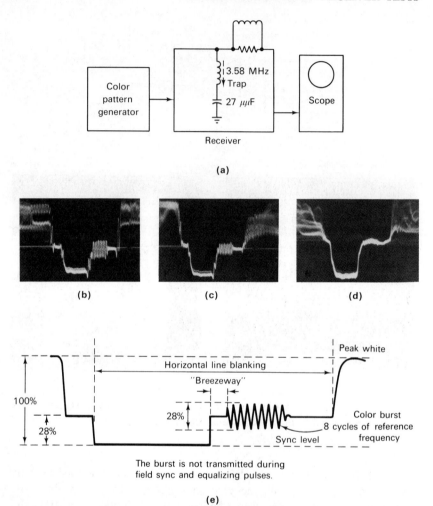

Figure 5–3 Subcarrier attenuation resulting from trap action: (a) test setup; (b) color burst with gain reduced 3.9 dB at 3.58 MHz; (c) color burst with gain reduced 6 dB at 3.58 MHz; (d) color burst with gain reduced 24 dB at 3.58 MHz; (e) standard waveform specifications.

always provides an additional attenuation of 6 dB, or the total attenuation of the color burst through the picture channel becomes 30 dB. Figure 5–4 shows comparative IF frequency responses for modern color receivers and old-model receivers.

Analysis of the color burst and other chroma signals requires that the oscilloscope provide a uniform vertical-amplifier response through

SECT. 5-1 / GENERAL CONSIDERATIONS

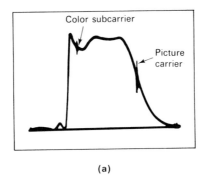

(a)

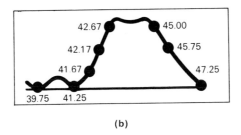

(b)

Intermediate Frequencies

	MHz
Lower adjacent channel sound IF	47.25
Picture carrier IF (accompanying channel)	45.75
Sound carrier IF (accompanying channel)	41.25
Color subcarrier	42.17
Higher adjacent channel picture IF	39.75

(c)

Figure 5-4 Key frequencies on IF frequency-response curves of old-model and modern color-TV receivers: (a) subcarrier falls on top of response curve in old-model receiver; (b) subcarrier falls at −6 dB on response curve of modern receiver; (c) key frequency relations in the IF channel.

3.58 MHz. It is also essential that the low-capacitance probe used with the oscilloscope be correctly adjusted for uniform frequency response. Otherwise, the deficiencies of the oscilloscope will be falsely charged to the receiver under test. An oscilloscope can be checked for frequency response by applying the output signal from a video sweep generator directly to the vertical-input terminals of the oscilloscope. Of course, the video sweep generator must provide a flat test signal (within ±1 dB). Similarly, a color pattern generator must provide standard horizontal sync pulses, with color bursts that have the same amplitude as the sync tips. Note that practically all color pattern generators provide maintenance adjustments for sync-pulse and color-burst amplitude.

When a color-bar video signal is passed through a color-subcarrier trap, the chroma component is removed from the Y signal, as illustrated in Fig. 5–5. Note in passing that this is an NTSC color-bar signal; it is widely used in laboratories, color-TV broadcast stations, and some service shops. An NTSC color-bar signal provides primary and complementary color bars at full saturation and brightness. On the other hand, most service shops utilize keyed-rainbow color-bar generators. A keyed-rainbow signal does not provide pure primary and complementary

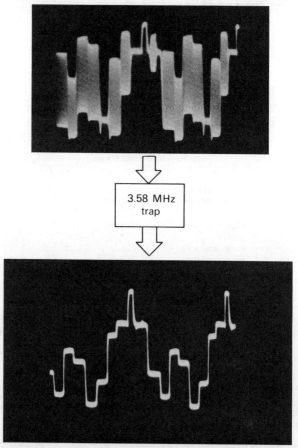

Figure 5–5 Subcarrier trap action removes chroma component from a color-bar signal.

colors, and the brightness and saturation vary from one color bar to the next. Nevertheless, technicians who have had extensive experience with both types of color-bar generators usually state that they can troubleshoot a color receiver equally well with either type of generator.

5-2 CHROMA BANDPASS AMPLIFIER TESTS

With reference to Fig. 5–1, the chroma bandpass amplifier is typically located in a branch circuit between the first and second video-amplifier stages. Functionally, a bandpass amplifier complements a Y amplifier in its filtering action. In other words, a bandpass amplifier passes the chroma component, and rejects the Y component of a color signal. As depicted in Fig. 5–6, a bandpass amplifier has a typical bandwidth of 1 MHz; the hump frequencies of the response curve are generally 3.1 and 4.1 MHz. A flat-topped frequency-response curve is exemplified in Fig. 5–6. However, some receivers employ a curve that slopes uphill from 3.1 to 4.1 MHz. This form of response is utilized to compensate for a downhill slope through the chroma-signal interval on the IF response curve. Note that the specified shape for the bandpass frequency response curve also depends upon the shape of the first video-amplifier response curve. Accordingly, the technician should consult the service data for the particular receiver when checking the bandpass-amplifier frequency response.

Normal circuit action for a bandpass amplifier is depicted in Fig. 5–7. In other words, when an NTSC color-bar signal is processed by the bandpass amplifier, the Y component is rejected, and the chroma component is passed. Thus, the bandpass has the opposite filtering function from a Y amplifier. Note that the color burst is passed with the chroma-bar signal in this example. However, some bandpass amplifiers are gated in such manner that the color burst is rejected. Observe that the color burst is passed by the bandpass amplifier in this example, because the bandpass stage is followed by the burst stage (Fig. 5–1). In any case, the receiver circuitry is arranged to blank out the color burst before the chroma signal arrives at the color picture tube. Unless the burst is blanked out, a yellowish-green horizontal-retrace pattern may become visible on the screen. Horizontal-retrace blanking also serves to gate out the burst.

If the bandpass amplifier is aligned with subnormal bandwidth, only comparatively low modulating frequencies can be processed. In turn, only the larger colored areas of the image can be reproduced. On the other hand, if the bandpass amplifier is aligned with abnormal bandwidth, an objectionable amount of the Y signal gains entry into the

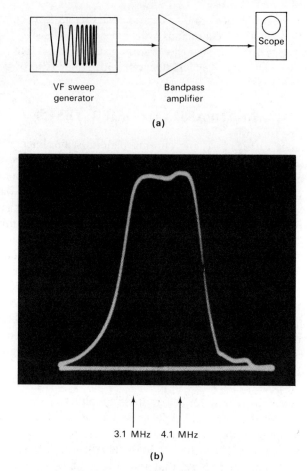

Figure 5–6 Check of bandpass amplifier frequency response: (a) test setup; (b) typical frequency-response curve.

chroma section. This interference distorts the reproduced hues, and produces annoying "line crawl" in the pattern. Thus, a 1-MHz bandwidth for the bandpass amplifier represents a practical compromise between reduction of color detail and passage of objectionable Y signal. Since the chroma signal is a double-sideband signal, a 1-MHz bandpass amplifier has an effective bandwidth of ±0.5 MHz. Inasmuch as the horizontal-scanning period is 63.5 μsec, this bandwidth permits the display of a colored interval ¼ inch wide. On the other hand, an effective bandwidth of ±0.25 MHz cannot provide display of a colored area less than ½ inch wide (if a 16-inch raster width is assumed). These bandwidth/color-bar relations are pictured in Fig. 5–8.

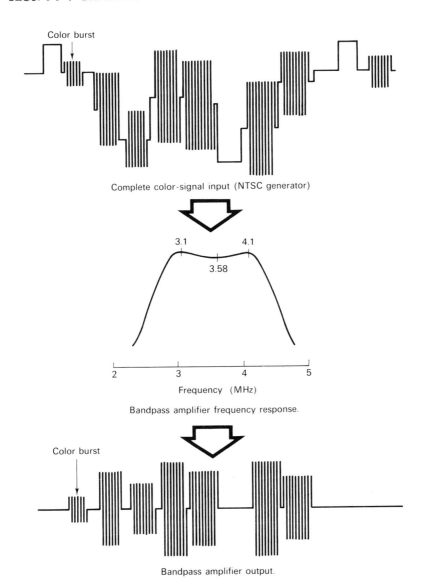

Figure 5–7 Bandpass amplifier filtering action.

5-3 BANDPASS-AMPLIFIER ALIGNMENT

Two basic methods of bandpass-amplifier alignment are utilized. One method employs a video-frequency sweep signal applied at the input of the bandpass amplifier, with a demodulator probe and oscilloscope

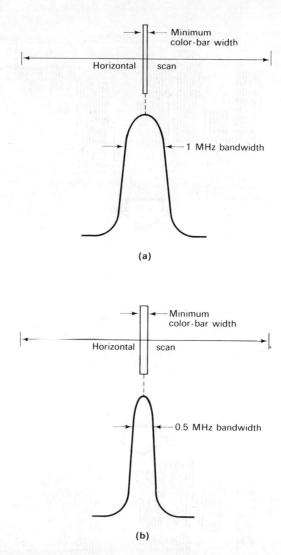

Figure 5–8 Bandwidth versus color-bar width relations: (a) minimum color-bar width for 1-MHz chroma bandwidth; (b) minimum color-bar width for 0.5-MHz chroma bandwidth.

connected at the output of the bandpass amplifier. This is a simple procedure that displays the frequency-response curve of the bandpass amplifier. Another alignment method uses an encoded sweep signal (video sweep modulated or VSM signal) applied at the input of the IF amplifier, with a demodulator probe and oscilloscope connected at the output of the bandpass amplifier. This VSM method is often preferred by technicians, because it shows how the IF amplifier, video amplifier, and bandpass amplifier operate together as a team. In other words, a VSM frequency-response curve displays the entire picture-channel chroma characteristic. A test setup as shown in Fig. 5-9 is utilized.

Note that the sweep/marker generator must be designed to provide a video sweep modulated IF signal. In other words, the generator forms a VSM IF signal by modulating a video-frequency sweep signal on the picture-IF carrier frequency (45.75 MHz). This encoded sweep signal passes through the IF amplifier and is demodulated through the video detector. Accordingly, a video-frequency sweep signal is applied to the video amplifier. Observe that this sweep signal has been varied in amplitude by the IF frequency-response characteristic. In turn, the sweep signal passes through the video amplifier and is further varied in amplitude by the video-amplifier frequency-response characteristic. Finally, the sweep signal passes through the bandpass amplifier, and it is again varied in amplitude by the bandpass-amplifier frequency-response characteristic. Thus, the frequency-response curve that is displayed on the oscilloscope screen shows the chroma-channel frequency response through the IF, video, and bandpass amplifier system. Either beat markers or absorption markers may be used.

5-4 BURST-AMPLIFIER TESTS

A burst amplifier can be compared with a chroma-bandpass amplifier in that it processes a 3.58-MHz signal. However, the bandwidth of a burst amplifier is comparatively small—typically 0.5 MHz. A burst amplifier is also gated, but in the opposite sense from that of a bandpass amplifier. In other words, a bandpass amplifier is often gated to reject the color burst, whereas a burst amplifier is gated to accept the color burst. Burst waveform aspects are shown in Fig. 5-10. An expanded burst display can be obtained only with a triggered-sweep oscilloscope. Note that the sweep speed employed in Fig. 5-10(c) is approximately 0.5 μsec/div. A normal burst display contains eight or nine cycles of subcarrier voltage. If the output from the burst amplifier has only three or four cycles, for example, it is indicated that the burst gating pulse is mistimed or has subnormal width. Also, the output waveform from the burst amplifier is checked for amplitude, as specified in the receiver service data.

Burst-amplifier tests may also be made to advantage with a dual-trace oscilloscope. For example, input/output waveform relations can

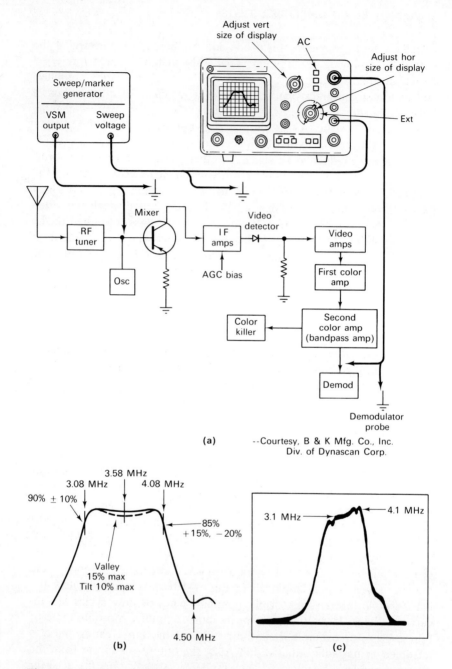

Figure 5–9 Video sweep modulation check of bandpass-amplifier frequency response: (a) test setup; (b) response curve showing tolerance limits, with beat markers; (c) response curve with absorption markers.

SECT. 5-4 / BURST-AMPLIFIER TESTS 139

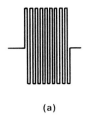

(a)

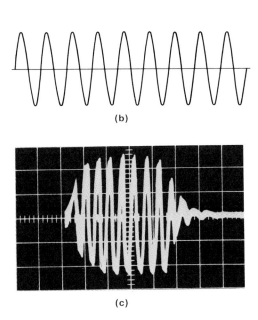

Figure 5–10 Burst waveform aspects: (a) ideal burst wavefront; (b) horizontal expansion of ideal burst; (c) typical display of expanded burst on scope screen.

be displayed as exemplified in Fig. 5–11. Note that some dual-trace oscilloscopes have comparatively limited display facilities, whereas others have elaborate facilities. Thus, practically all dual-trace oscilloscopes have independent vertical-gain controls, so that the vertical deflection of each waveform can be adjusted individually. On the other hand, a dual-trace oscilloscope may or may not provide individual times bases for the two waveforms. When a single time base is used for both displays, both waveforms must be expanded simultaneously. On the other hand, if separate times bases are provided, the upper display in Fig. 5–11 can be expanded as illustrated in Fig. 5–10(c) without expansion of the lower waveform. One of the most important advantages of a dual-trace oscilloscope is its ability to indicate the timing of two waveforms. For example, a dual-trace oscilloscope can display a gating pulse and a chroma waveform simultaneously, and shows directly whether the pulse is correctly timed with respect to the burst, for example.

Burst amplification is essentially linear in most cases, as exemplified in Fig. 5–11(b). However, owing to component or device defects, or to incorrect bias-supply voltage, burst amplification may become nonlinear, as illustrated in Fig. 5–12(b). In this example, the stage is operating in class B, rather than in class A. Nonlinear amplification, of itself, has little effect on the tightness of color-sync lock, provided that the peak-to-peak voltage of the output waveform is normal. This is

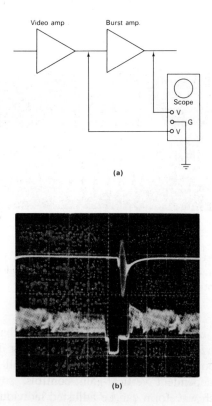

Figure 5–11 Example of a dual-trace oscilloscopic display: (a) test setup; (b) normal display.

SECT. 5-4 / BURST-AMPLIFIER TESTS

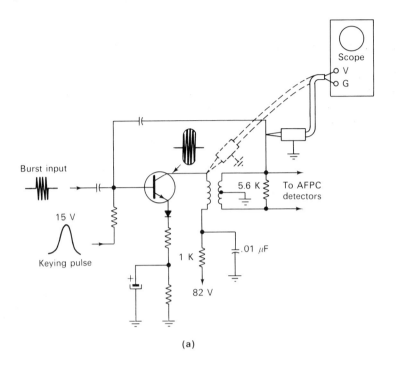

(a)

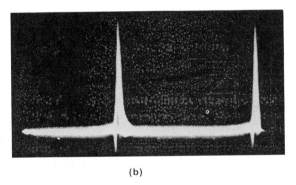

(b)

Figure 5–12 Output waveform from burst amplifier is sometimes distorted: (a) test setup; (b) nonlinear output waveform owing to amplifier malfunction.

because nonlinear amplification generates even harmonics that are rejected by the following tuned circuits. In turn, the signal that is applied to the color-sync section is a 3.58-MHz sine wave, regardless of nonlinear distortion that may occur in the burst-amplifier stage.

5-5 SUBCARRIER-OSCILLATOR CHECKOUT

Many subcarrier oscillators employ shock-excited quartz-crystal action. In other words, the color-burst waveform is applied to a 3.58-MHz crystal, which in turn "rings" in phase with the burst signal. A typical configuration is shown in Fig. 5–13. Observe that the normal output waveform has a slow decay, and is approximately a CW voltage. There

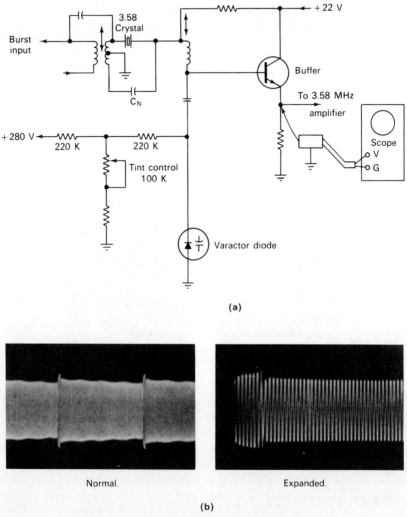

Figure 5–13 Check of subcarrier oscillator circuit action: (a) typical configuration; (b) normal output waveform.

SECT. 5-6 / CHROMA DEMODULATOR CHECKOUT

is comparatively little capacitive feedthrough from the input to the output of the crystal holder, owing to the neutralizing capacitor C_N. This neutralizing capacitor also has an effect on the ringing phase of the crystal. If the neutralizing capacitor becomes "open," the crystal phase will shift and there will be an abnormal amount of color-burst feedthrough. On the other hand, if the neutralizing capacitor becomes leaky, the decay rate of the ringing waveform will increase. In severe situations, the color intensity varies objectionably across the picture-tube screen from left to right. Note that the tint control in Fig. 5–13 operates by variation of capacitance across the crystal, thereby varying its ringing phase.

Another widely used subcarrier-oscillator configuration is the AFPC design, shown in Fig. 5–14. This arrangement employs a free-running 3.579545-MHz oscillator that is controlled in frequency and phase by the output control voltage from a phase detector. This phase detector is essentially a discriminator that compares the color-burst phase with the phase of the subcarrier-oscillator output voltage. A reactance device, such as a varactor diode, is utilized to shunt an adjustable value of capacitance across the oscillator tank. In turn, the oscillator is normally held on frequency and precisely in phase with the color burst. When malfunction occurs, the output amplitude from the oscillator may be subnormal and/or the waveshape may be seriously distorted. These trouble conditions are shown by the oscilloscope. Again, the output amplitude may be normal, and the waveshape may be acceptable, but the oscillating frequency may be in substantial error. Since an oscilloscope cannot measure this frequency to a very high degree of precision, it is advisable to connect a lab-type digital frequency counter at the output of the 3.58-MHz amplifier, as depicted in Fig. 5–14.

5-6 CHROMA DEMODULATOR CHECKOUT

Chroma demodulators are usually checked with a keyed-rainbow color-bar signal. With reference to Fig. 5–15, this signal comprises 11 color bursts at black level between successive horizontal-sync pulses. Each burst consists of approximately a dozen cycles of 3.563795-MHz sine-wave voltage, commonly referred to as 3.56 MHz. Other terms of this type of chroma signal are offset color subcarrier, sidelock signal, and linear phase sweep. Note that the first burst following a horizontal sync pulse is employed for color synchronization by the receiver system. In turn, the ten following bursts are processed by the chroma demodulators in the receiver, and are displayed as ten vertical color bars on the picture-tube screen. As indicated in the diagram, the bursts follow one another

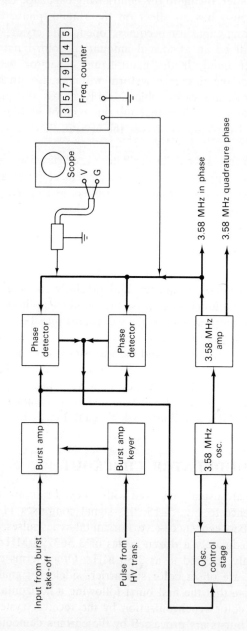

Figure 5-14 Oscilloscope checks waveshape and amplitude; digital frequency counter checks frequency.

144

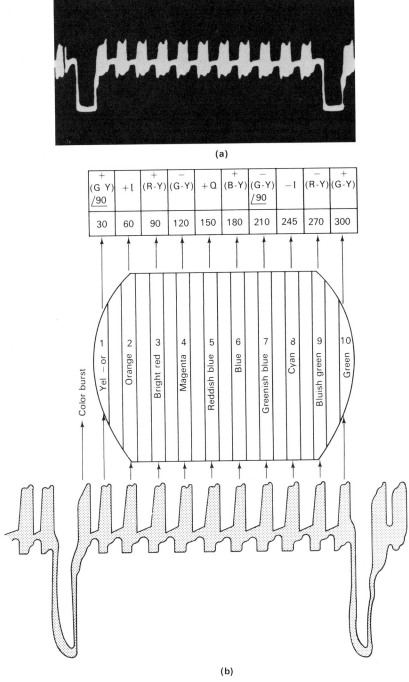

Figure 5–15 Keyed-rainbow color-bar signal characteristics: (a) typical oscilloscope display; (b) color-bar phase and hue identifications.

at 30-deg intervals. In other words, each burst is advanced 30 deg in phase with respect to the preceding burst. This phase progression results in a display of various hues, as noted in the diagram. Each of the burst phases has an identifying term, such as R-Y, B-Y, G-Y, I, Q, and so on.

A chroma demodulator is a phase detector and also an amplitude detector. Phase demodulation occurs with respect to the subcarrier-oscillator phase. In other words, if the incoming chroma signal is in phase with the subcarrier voltage, maximum demodulated output is obtained. On the other hand, if the incoming chroma signal is 90 deg out of phase with the subcarrier voltage, zero demodulated output is obtained. Again, if the incoming chroma signal varies in amplitude at any given phase, the output amplitude will vary proportionally at that phase. When a keyed-rainbow signal is applied to a color receiver and an oscilloscope is connected at the chroma-demodulator outputs, as exemplified in Fig. 5–16, waveforms are normally displayed as depicted in the diagram. Observe that the R-Y waveform crosses the zero axis at bar 6, the B-Y waveform crosses at bars 3 and 9, and the G-Y waveform crosses at bars 1 and 7.

In practice, chroma-demodulator waveforms have an appearance as exemplified in Fig. 5–17. Owing to limited bandwidth, the demodulator output waveforms do not have square corners; they have comparatively slow rise and rounded tops. Observe that the base lines of the waveforms in Fig. 5–17(a) and (b) are comparatively straight. Substantially curved base lines are trouble symptoms. Crossovers are approximately correct in these examples. Note that all crossovers will shift if the hue control is readjusted. The important consideration is that if the hue control is adjusted to make the R-Y waveform crossovers correct, the B-Y and G-Y waveform crossovers will also be correct. In the event of incorrect crossovers, it is indicated that there is a phasing error in the subcarrier injection voltage. For example, if incorrect crossovers are observed, the technician would readjust the slug in T1 (Fig. 5–16). Then, if the phasing error persists, it is probable that there is a defective capacitor or coil in the T1 and T2 network. Or a coupling capacitor to the malfunctioning demodulator may be "open" or leaky. Sometimes a faulty demodulator diode simulates a phasing error trouble symptom.

Next, observe the relative R-Y, B-Y, and G-Y signal amplitudes depicted in Fig. 5–17(c). These are the correct chroma-output waveform ratios for a representative receiver. Note that these ratios may not necessarily be displayed at the chroma-demodulator outputs—the following R-Y, B-Y, and G-Y amplifiers may provide unequal gains to develop the correct chroma-output waveform ratios. In other words, the ratios specified in the receiver service data will always be observed at the picture-tube input circuit in normal operation. Gain controls are

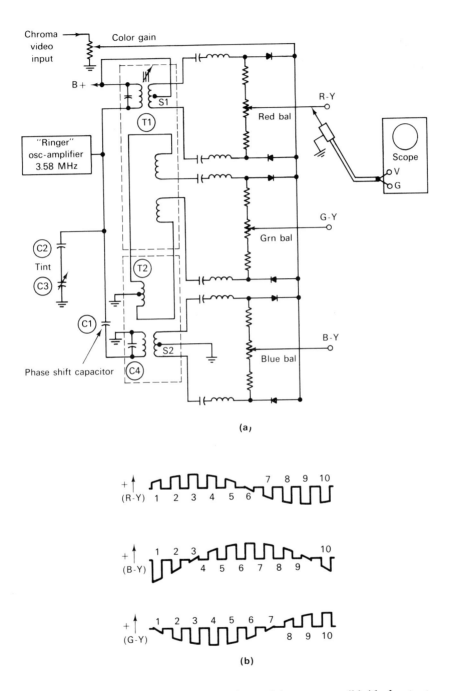

Figure 5–16 Chroma demodulator checkout: (a) test setup; (b) ideal output waveforms, keyed-rainbow signal.

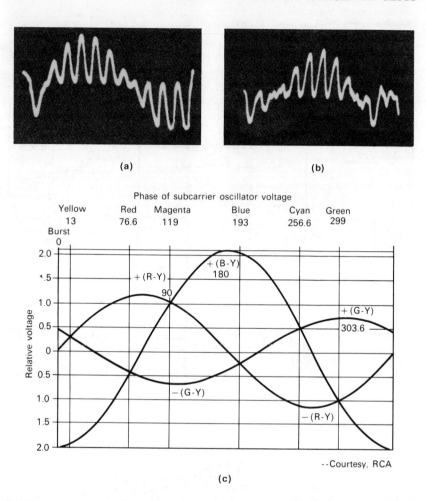

Figure 5–17 Chroma demodulator waveform characteristics: (a) typical R-Y demodulator output; (b) typical B-Y demodulator output; (c) relative output amplitudes for a representative receiver. (*Courtesy of* RCA.)

provided in many receivers for setting the R-Y, B-Y, and G-Y ratios. If the specified signal ratios cannot be obtained by adjusting the gain controls, it is indicated that there is a component or device defect in one of the channels. Localization is generally accomplished by chroma signal-tracing procedures. In turn, the faulty component or device can usually be pinpointed by means of DC voltage and resistance measurements.

When a video-frequency sweep signal is applied at the input of a chroma-demodulator stage, and a demodulator probe with an oscillo-

SECT. 5-6 / CHROMA DEMODULATOR CHECKOUT

scope is connected at the output of the stage, a frequency-response curve is normally displayed as exemplified in Fig. 5–18. A chroma-demodulator response curve is somewhat similar to a bandpass-amplifier response curve, except that the former has approximately one-half the bandwidth of the latter. In other words, a chroma-demodulator response curve typically has a 0.5-MHz bandwidth. Note that the curve can be marked either with beat markers or with absorption markers, depending upon the sweep-generator facilities. Absorption markers are preferred by many technicians, because beat markers are often accompanied by spurious markers that may be confusing. Spurious markers result from beats between marker harmonics and sweep harmonics. Absorption markers are produced by passive traps. In turn, there are no harmonics associated with an absorption marker, and spurious absorption markers do not occur.

If a sweep-frequency generator does not have built-in absorption-marker facilities, an external absorption-marker box can be employed, as shown in Fig. 5–19. A typical marker box provides check frequencies of 0.5, 1.5, 3.1, 3.58, 4.1, and 4.5 MHz. Markers are displayed simultaneously on a frequency-response curve. A terminal board is provided on the marker box so that the technician can touch his finger to a selected terminal. Each terminal corresponds to one of the foregoing

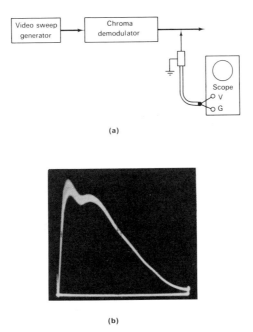

Figure 5–18 Typical R-Y demodulator frequency-response curve: (a) test setup; (b) frequency-response curve.

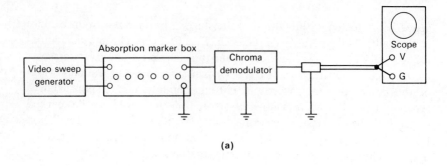

(a)

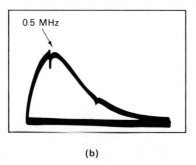

(b)

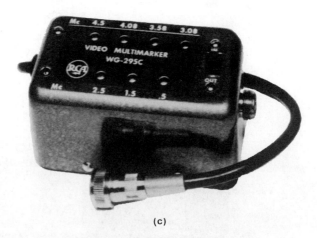

(c)

Figure 5–19 Application of external absorption-marker box: (a) test setup; (b) chroma-demodulator response curve with absorption markers; (c) an absorption marker box. (*Courtesy of* RCA)

SECT. 5-7 / SPECIALIZED COLOR-TV TEST SIGNALS

check frequencies. When the terminal is touched, the corresponding marker trap is slightly detuned, and the marker moves visibly along the curve. This reaction enables the technician to identify the frequency of any marker displayed on a response curve. Note that an alignment system that includes a post-marker indicator arrangement avoids the difficulty of spurious beat markers.

5-7 SPECIALIZED COLOR-TV TEST SIGNALS

Several specialized color-TV test signals are available, both for transmitter adjustment and for receiver checkout. The most familiar of these is the special color-bar pattern that is transmitted by various TV stations, usually in the early morning hours. As pictured in Fig. 5–20, this color-

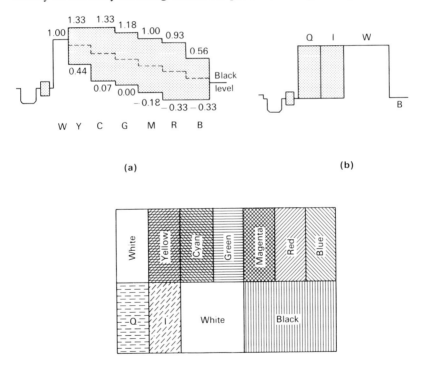

Figure 5–20 Typical color-bar pattern transmitted by a TV station: (a) video signal, first half of vertical field; (b) video signal, second half of vertical field.

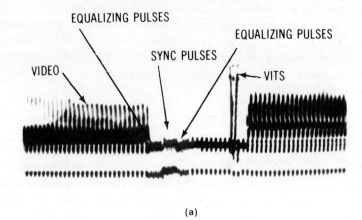

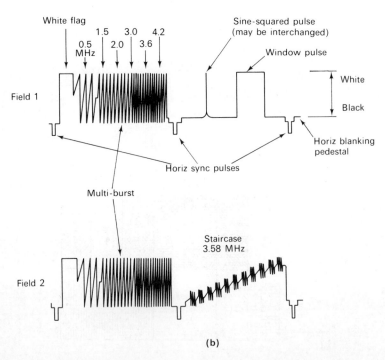

Figure 5–21 VITS waveforms: (a) display of the VITS waveform in the vertical-sync interval; (b) detail of expanded VITS signal.

bar signal and pattern are somewhat more elaborate than those provided by conventional NTSC color-bar generators. Note that two signal waveforms are used, each of which is active over one-half of the vertical field. In turn, a split-field pattern is displayed. The upper pattern is basically a familiar NTSC color-bar pattern that displays the primary and complementary hues, plus a reference white level. On the other hand, the lower pattern is a chroma-bar pattern that displays the I and Q hues, plus reference white and black levels. This color-bar pattern is useful both for transmitter and for receiver adjustments. However, a service shop requires the availability of a color-bar generator for bench work.

Another specialized color-TV test signal is transmitted during the vertical-retrace interval in color programs. This is called a vertical-interval test signal (VITS waveform), as illustrated in Fig. 5–21. It comprises a multiburst signal, a 3.58-MHz staircase signal, a window pulse, and a sine-squared pulse. The multiburst signal provides check frequencies of 0.5, 1.5, 2.0, 3.0, 3.6, and 4.2 MHz; it is equivalent to a video-frequency sweep signal. As noted previously, a staircase signal is used chiefly to check a system for amplitude linearity. The window pulse and sine-squared pulse are employed to check transmitter system response; a sine-squared pulse provides a critical transient-response test. A VITS waveform is used by experienced technicians to check picture-channel characteristics in color receivers.

Still another specialized color-TV test signal that is transmitted during the vertical-retrace interval in color programs is the vertical interval reference signal (VIR signal). Its purpose is to reduce undesirable variations in hue at the color receiver. This is accomplished by providing a reference chroma signal that assists transmitter system personnel in adjustment of various signal parameters so that all programs are transmitted with precise amplitude and phase characteristics. Note that a VIR signal supplements, but does not replace, the VIT signal. As shown in Fig. 5–22, the VIR signal has the burst (color-subcarrier) phase, and is placed on a 70 percent Y level. This particular Y level was specified because it is about the same luminance level as skin tones in typical programming. As noted previously, skin or flesh tones are the most critical hues in a color-TV image. The VIR signal is inserted into line 19 of both fields, whereas the VIT signal is inserted into line 18 of both fields.

5-8 SCR SWEEP-CIRCUIT WAVEFORMS

Current waveforms are of basic importance in checking the operation of silicon controlled rectifier sweep circuits. A basic configuration is shown in Fig. 5–23. To display a current waveform, a current probe should be utilized. In the case of a dual-trace oscilloscope, two current

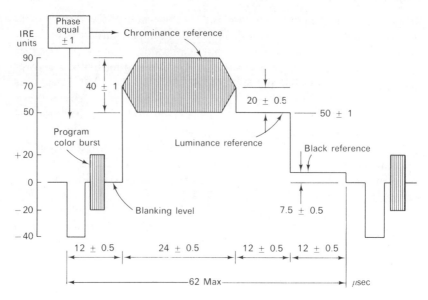

Figure 5–22 Specifications for the VIR signal.

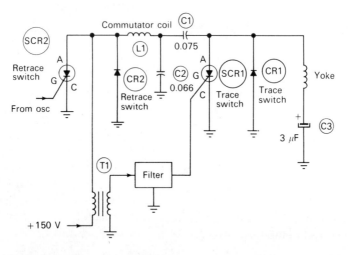

Figure 5–23 Basic SCR horizontal-deflection configuration.

probes should be employed. Key current waveforms for the sweep circuit are shown in Fig. 5–24, arranged in normal phases. These are the waveforms that should be observed when they are checked two at a time with a dual-trace oscilloscope. Note that horizontal flyback (retrace) occurs between T4 and T0. Because of circuit interaction, it is

SECT. 5-8 / SCR SWEEP-CIRCUIT WAVEFORMS

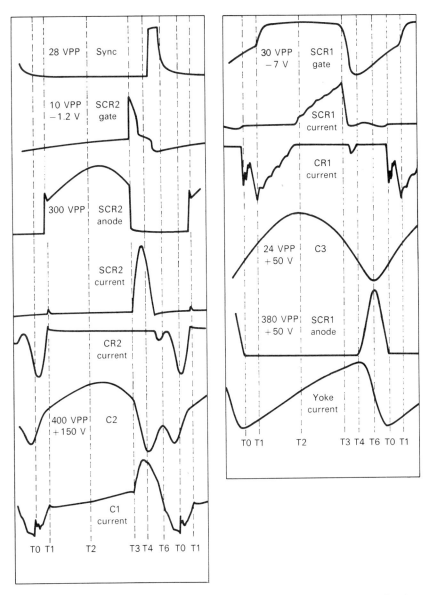

Figure 5–24 Key current waveforms for an SCR horizontal-deflection system.

not always possible to pinpoint a defective component or device on the basis of waveform distortion. However, preliminary analysis of a trouble symptom may be facilitated by waveform checks.

6

INDUSTRIAL-ELECTRONICS CIRCUIT ACTIONS AND WAVEFORMS

6-1 GENERAL CONSIDERATIONS

Saturable reactors (saturable inductors) are used in various types of industrial-electronics circuits. A saturable inductor is a nonlinear inductor in which a comparatively small value of current produces magnetic core saturation. In turn, the nonlinear coil has an inductance value that depends upon the amplitude of the current waveform flowing through its winding. Figure 6-1 shows the basic relation between these parameters. Thus, if a large-amplitude sine-wave current flows through the winding, the inductance of the coil changes abruptly as the current amplitude exceeds the saturation value of the core. Note that while the current amplitude is between limits A and B, the core is unsaturated and the effective inductance of the coil is high, as indicated by L_u. On the other hand, when the current amplitude exceeds the value B in the positive direction or value A in the negative direction, the core saturates and the effective inductance of the coil drops to a very low value, as at L_s.

Observe the magnetization curve depicted in Fig. 6-1. It shows an ideal graph of flux density in the core versus current amplitude in the winding. It shows that the flux density in the core is proportional to the current that flows through the coil while the current amplitude is between limits A and B, and the inductance value is then equal to L_u. During the

157

158 CHAP. 6 / INDUSTRIAL-ELECTRONICS CIRCUIT ACTIONS

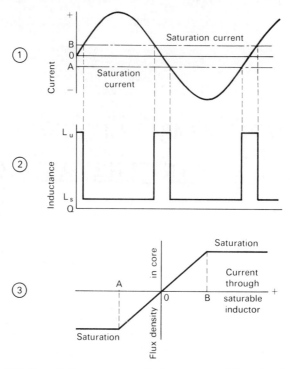

Figure 6-1 Relations between current, inductance, and flux-density values in a saturable inductor.

time that the current is more positive than B, or more negative than A, the core is saturated, and the inductance is then equal to the low value L_s. An example of a circuit in which a saturable inductor is used to form a pulse waveform output from a sine-wave input is shown in Fig. 6-2. Note that the series circuit L_1C_1 and L_2 is nearly resonant to the sine-wave frequency. Also, the value of inductor L_1 is chosen so that the circuit is slightly inductive when L_2 is unsaturated, and slightly capacitive when L_2 is saturated. Thus, while L_2 is saturated, the circuit current leads the applied voltage by nearly 90 deg.

Observe that since the voltage across L_2 in Fig. 6-2 leads the current by nearly 90 deg, the voltage across L_2 is nearly 180 deg out of phase with the applied voltage; because the saturated inductance L_s (Fig. 6-1) of L_2 is almost zero, the voltage across the inductor has a very small amplitude, as shown by curve S (Fig. 6-2). If L_2 is replaced by an ordinary coil with an inductance equal to that of the unsaturated

SECT. 6-1 / GENERAL CONSIDERATIONS 159

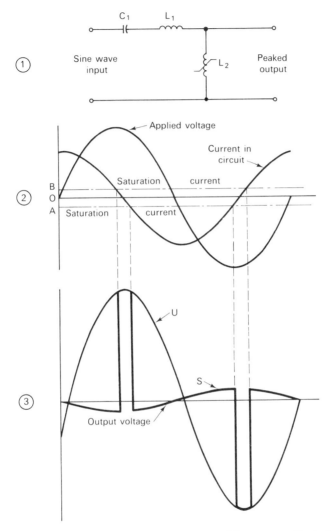

Figure 6-2 Basic peaking-circuit action provided by a saturable inductor.

inductance of L_2 (L_u in Fig. 6–1), the circuit will become inductive and the current will lag the applied voltage by nearly 90 deg. In this situation, the voltage across L_2 is in phase with the applied voltage, but because the circuit is near series resonance and because the reactance of L_2 is now large, a large-amplitude sine wave such as U in Fig. 6–2 will appear across L_2. Since L_2 is saturated during most of the cycle, the

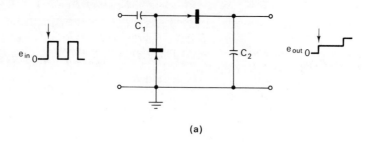

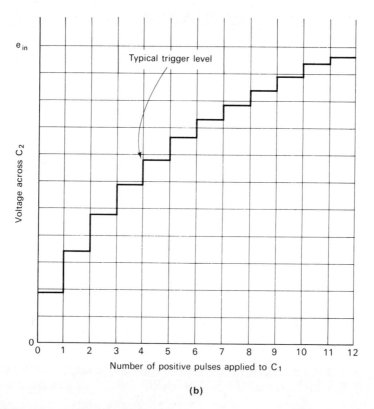

Figure 6-3 A basic step-by-step counter circuit: (a) counting configuration; (b) output waveform.

voltage across this inductor is essentially the small-amplitude sine wave S. However, while the current passes from the saturation value A through zero to the saturation value B in the opposite direction, the core is unsaturated. During this short interval, the voltage across L_2 changes abruptly from waveform S to waveform U, producing a high-amplitude pulse. When the coil again becomes saturated, the voltage across L_2 drops back to waveform S. Note that the duration of the pulse that is produced is made short by the series-resonant circuit, which causes a large current to flow through L_2 so that only a very short time is required for the current to pass from the saturation value in one direction to the saturation value in the other direction.

6-2 STEP-BY-STEP COUNTING CIRCUIT WAVEFORMS

Various types of counting circuits are encountered in industrial-electronics equipment. One of the simplest step-by-step counting configurations is shown in Fig. 6–3. It utilizes a pulse-waveform input and produces an output stair-step interval for each applied pulse. Observe that these steps decrease in amplitude exponentially as the voltage across C_2 approaches its final value. This counting circuit may be employed as a frequency divider, and used to trigger a following silicon controlled rectifier, for example. When the stair-step waveform reaches the trigger voltage, the utilization circuit suddenly discharges C_2, and the counting sequence starts over again. A control is usually provided to adjust the trigger point. Thereby, the frequency-divider "count" can be changed, and the time interval between triggers set as required. This arrangement is often called an *integrating counter*.

6-3 PHASE-SHIFTER WAVEFORMS

Phase-shifting circuits are used extensively in subsections of industrial-electronics systems. An L-section RC phase shifter is shown in Fig. 6–4. Observe that an AC voltage applied to this circuit causes a current to flow that has an amplitude which is determined by the impedance of the circuit. Since the impedance is capacitive, the current i leads the applied voltage e by an angle θ, which is equal to 60 deg in the example of Fig. 6–4. In turn, the voltage drop e_R that occurs across resistor R is in phase with the current that flows through it. Accordingly, e_R must lead the impressed voltage by an angle θ. If the output from this L section is applied to a second similar phase shifter, the phase of the output voltage is shifted by an angle θ again. Thus, the output waveform from this

162   CHAP. 6 / INDUSTRIAL-ELECTRONICS CIRCUIT ACTIONS

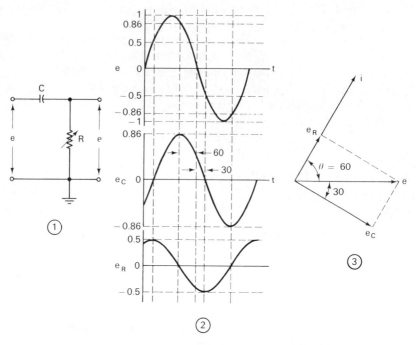

Figure 6–4 Circuit action of L-section RC phase shifter.

second phase shifter leads the first input-voltage waveform by an angle of 120 deg.

Note that if the resistance of R is varied in the circuit of Fig. 6–4, the phase angle of the current flowing in the circuit is also varied. However, if the value of R were reduced to zero, there would be no voltage drop across R, and the circuit would not function. In other words, it is impractical to obtain a 90-deg phase shift with a single L section. When a phase shift of 180 deg is required, three L sections are connected in series. Note that since the reactance of a capacitor varies with frequency, a three-section L phase shifter provides a 180-deg shift at only one frequency. Waveforms in a three section RC phase shifter are exemplified in Fig. 6–5. In this circuit, each section provides a 60-deg phase shift at the specified operating frequency. Observe that each section also attenuates the output waveform. A dual-trace oscilloscope test setup for checking phase shift and waveform amplitudes is depicted in Fig. 6–6.

If a single-trace oscilloscope is utilized, phase shift can be measured by external synchronization of the time base section, as shown in Fig.

SECT. 6-3 / PHASE-SHIFTER WAVEFORMS

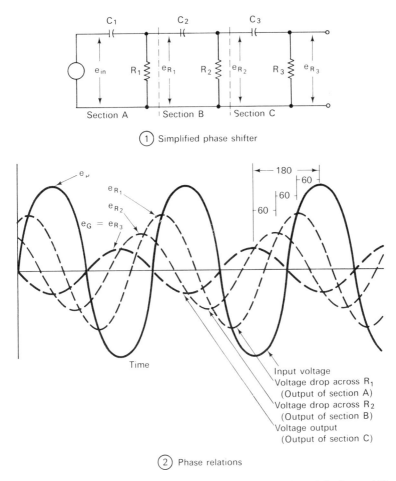

Figure 6-5 Waveform relations in an L-type three-section RC phase shifter.

6-7. In this example, the time base is synchronized by the output waveform from the phase shifter. In turn, the horizontal sweep will always start at the same time, regardless of the vertical-input signal. Therefore, when the vertical-input lead of the oscilloscope is applied at the input of the second section, a reference sine-wave display is obtained. Next, the vertical-input lead is applied at the output of the second section, and the phase of the sine-wave display is compared with that of the reference waveform. This phase difference is equal to the phase shift that is provided by the second section. If the vertical-input lead is then

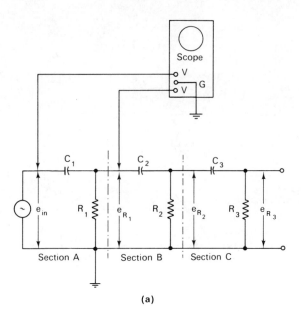

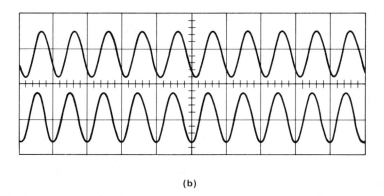

Figure 6–6 Phase measurement with a dual-trace oscilloscope: (a) test setup; (b) dual traces display a phase shift.

moved to the input of the first section, the input phase can be compared with the reference output phase. This phase difference is equal to the total phase shift of the three sections. It is evident that the chief distinction between this method and the dual-trace method of phase measurement is that a sequential test procedure is utilized instead of a simultaneous test procedure.

SECT. 6-4 / ELIMINATION OF TRANSIENT IN AC WAVEFORM

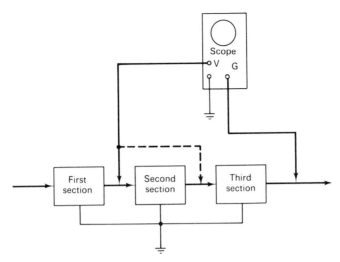

Figure 6–7 Phase measurement by use of external synchronization.

6-4 ELIMINATION OF TRANSIENT IN SWITCHED AC WAVEFORM

Industrial-electronics equipment, typified by welding units, requires close control of the power content in switched-AC waveforms consisting of a predetermined number of cycles. In other words, when a sine waveform is suddenly switched into a load, the resulting load current may or may not contain a transient component in addition to the sine-wave component. If the load current draws a transient component, the power developed in the load will be greater than if the current waveform consists of a sine wave only. Thus, three cycles of sine-wave voltage may be switched into a load, but the power that develops will depend on whether these three sine-wave cycles are or are not accompanied by an exponential transient flow of current. The additional power developed by the exponential current flow depends upon the amplitude of the transient. As will be explained, the transient component will be eliminated if the sine-wave voltage is applied to the load at a certain precise point in the voltage waveform.

With reference to Fig. 6–8, a moderately reactive load is under consideration. This load might be inductive or capacitive. In this example, a somewhat inductive load is assumed. As shown in the waveforms, a large transient current will be drawn in addition to a sine-wave current if the load is switched into the line as the voltage waveform goes

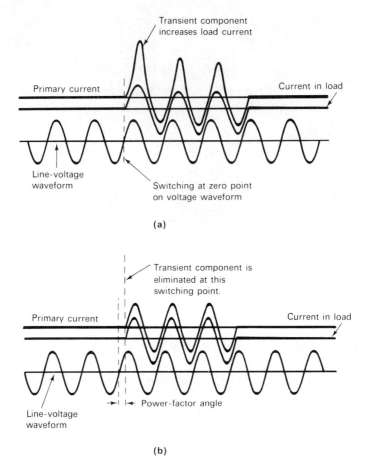

Figure 6–8 Example of phase relations required to suppress a transient current in an inductive load: (a) line voltage switched into load at zero phase angle; (b) line voltage switched into load at the load's phase angle.

through zero. On the other hand, if the load is switched into the line after the voltage waveform goes through zero, it is possible to suppress the transient current, although the sine-wave current is the same as before. Observe that the transient current will be suppressed if the load is switched into the line by a lagging number of degrees equal to the power-factor angle of the load. For example, if the power-factor angle of the load were 60 deg, then the line voltage would be switched into the load 60 deg after it crossed the zero axis. This is another example of the basic significance of phase relations in various types of industrial-electronics equipment.

6-5 PLASMA OSCILLATION WAVEFORM

Radio frequency interference to radio receivers and television receivers is sometimes caused by industrial-electronics equipment. Aside from sparking contacts and radiation escaping from high-frequency heating units, plasma oscillation can be a source of interference. As exemplified in Fig. 6–9, a simple thyratron pulsing circuit has a load voltage waveform that contains a strong plasma-oscillation component. Suppression

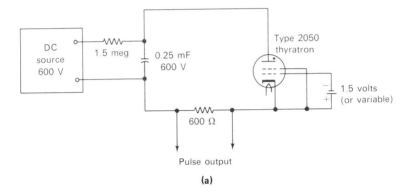

(a)

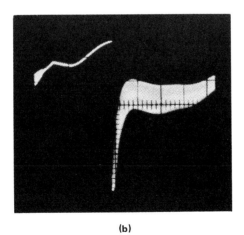

(b)

Figure 6–9 RF interference is caused by plasma oscillation: (a) typical test circuit; (b) output waveform showing plasma oscillation.

of the RF interference requires shielding of the pulse generator. Small RF chokes in series with the leads to the load also assist in minimizing high-frequency radiation. In some cases, it is also necessary to insert an RF filter into the 60-Hz leads of the power supply.

6-6 CLAMPING CIRCUIT WAVEFORMS

Clamping circuits are encountered in various types of industrial-electronics equipment. For example, CRT's are used in process-control units in which the displayed trace is divided into precisely equal time intervals. To ensure that the start of the trace always coincides with the zero-reference point on the screen, a clamping circuit is ordinarily utilized. A basic clamping circuit is shown in Fig. 6–10. Note that the CRT is deflected by push-pull sawtooth waveforms. At the end of the sweep, the beam retraces from B to A. In turn, the clamping circuit forces point A to remain fixed despite variations in amplitude of the sawtooth deflection waveform. If diodes D1 and D2 were not connected, and R3 were set to its $+$ 50 volt point, the average potential of plate P1 would be 50 volts more positive than P2, causing the beam to be attracted to the left of center in the absence of sweep voltage. When the sawtooth voltage waveforms are applied, the potentials of the plates swing equal amounts above and below the average values. Thus, any variation in the amplitude of the sweep voltages would cause the start of the sweep to change position on the screen.

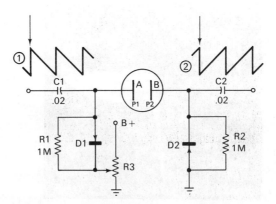

Figure 6–10 Basic clamping circuit.

SECT. 6-6 / CLAMPING CIRCUIT WAVEFORMS

To clamp the starting point of the sweep (point A, Fig. 6–11) to a fixed potential, diodes D1 and D2 are connected in the circuit. The first cycle of the sawtooth voltage (Fig. 6–11) is shown before the diodes are connected, so that the change brought about by the diode action is apparent. Before the diode was connected, the voltage at the plate of P1 varied between $+100$ volts and ground potential. However, when the diode is connected into the circuit, and the arm on R3 is set so that the cathode of the diode is at $+50$ volts, any rise of plate voltage above this value is short-circuited by conduction of D1. When the diode conducts, capacitor C1 is rapidly charged, so that the voltage on plate P1 drops quickly to $+150$ volts, and thereafter the voltage falls in accordance with the sawtooth waveform. Thus, the effect of the diode is to put a charge on C1 that is sufficient to change the average potential of the sawtooth voltage from $+150$ volts to $+100$ volts. Note that the small amount of charge that leaks from C1 during the forward-trace interval is replenished at retrace, so that the starting point of the sweep is held

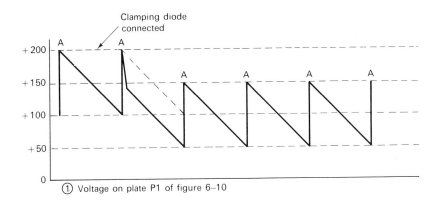

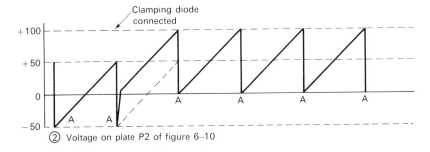

Figure 6–11 Sawtooth voltage waveforms with clamped sweep.

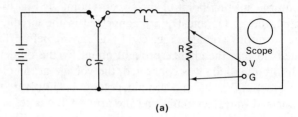

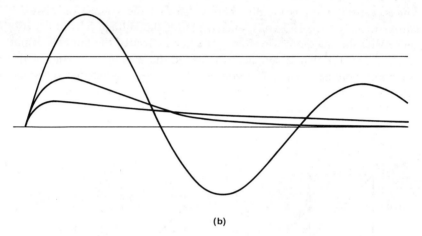

Figure 6-12 Overdamped, critically damped, and underdamped LCR circuit waveforms. Lowest amplitude: overdamped; medium amplitude: critically damped; highest amplitude: underdamped; horizontal line: applied DC level.

fixed at +150 volts. Consequently, any variations in amplitude of the sweep-voltage waveform can affect only the length of the sweep, and the starting point remains constant.

In a similar manner, the voltage to which C2 is charged in Fig. 6-11 is changed from zero to + 50 volts by the action of diode D2. In turn, the starting potential of each sweep on plate P2 is clamped to ground potential. Observe that the difference between the average potentials of the two plates is the same as with the diodes disconnected, so that the centering action is the same, but that the voltage to which the arm on R3 must be set to maintain this condition is higher in the case when the clamping diodes are used.

6-7 OVERDAMPED, CRITICALLY DAMPED, AND UNDERDAMPED WAVEFORMS

Some industrial-electronics units switch a charged capacitor into the load. A basic arrangement is shown in Fig. 6–12. In most cases, the load circuit is somewhat inductive. In turn, three modes of discharge may occur, depending upon the circuit parameters. If the resistance value is comparatively low, the circuit will be underdamped and the current waveform will have a damped sinusoidal waveshape. This is the oscillatory mode of operation. Next, if the resistance value is increased to the point that the oscillatory mode is just suppressed, the circuit will be critically damped. In the critically damped mode of operation, the current attains its maximum possible positive-peak value, without a subsequent reversal of current flow. Finally, if the resistance value is increased beyond the critical point, the circuit will be underdamped. In the underdamped mode of operation, a single polarity of surge current flows as before, except that its peak amplitude is less. An oscilloscope is practically indispensable in checking the mode of operation.

6-8 DELAY TIME MEASUREMENT WITH DUAL TRACE OSCILLOSCOPE

Control processes in high-speed operations are often dependent upon rapid switching action of transistors. In other words, if a component or device malfunction slows down the switching time of an operant transistor, a trouble symptom occurs. To check the switching time of a transistor, a dual-trace oscilloscope with a triggered and calibrated time base is most useful. The vertical amplifiers must have ample frequency response, so that the leading edges of displayed pulses are not distorted. A typical test setup and normal input/output waveforms are shown in Fig. 6–13. In most applications, the turn-on time is of most importance. Some applications also require a rapid turn-off time. Normal turn-on and turn-off times are usually noted in the equipment servicing manual. In case of doubt, a comparison test can sometimes be made on a similar equipment installation that is operating satisfactorily.

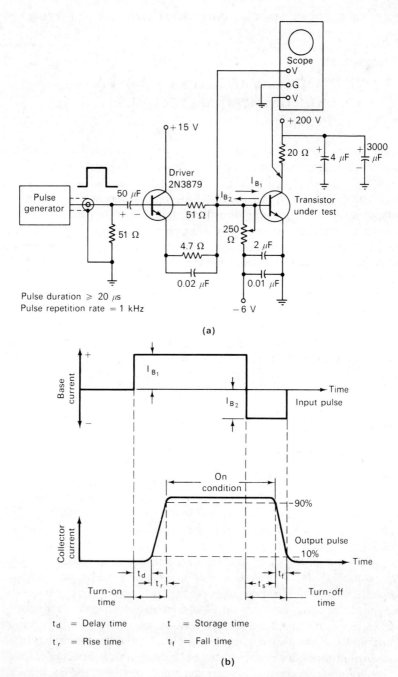

Figure 6–13 Test setup for checking transistor switching time: (a) configuration; (b) typical input/output waveforms.

6-9 INVERTER UNIT WAVEFORM CHECKS

Switching action is employed in transistors for power inverters, as exemplified in Fig. 6–14. This is a typical industrial-electronics power-supply unit that provides up to 100 watts output at a frequency of approximately 18 kHz. It employs a push-pull oscillator transformer that operates as a saturable inductor. However, the output transformer functions in a nonsaturated mode. Positive feedback takes place from the collectors of the transistors to the primary of the oscillator trans-

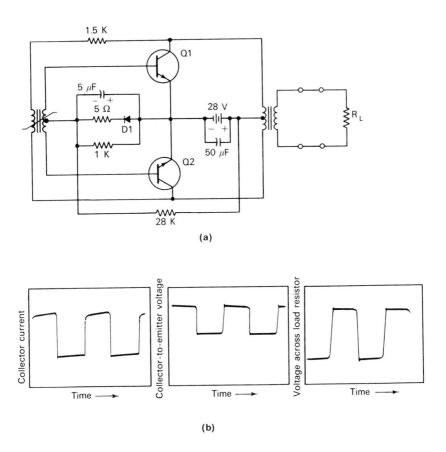

Figure 6–14 A power inverter arrangement: (a) circuit and test setup; (b) normal operating waveforms.

174 CHAP. 6 / INDUSTRIAL-ELECTRONICS CIRCUIT ACTIONS

former. Note that the transistors are driven into saturation immediately after the core of the oscillator transformer saturates; the transistors are alternately saturated and cut off. If the power output becomes subnormal, a check should be made of the oscillating frequency. In other words, the power output decreases if the oscillating frequency is subnormal.

6-10 FREQUENCY CONVERTER WAVEFORMS

Industrial plants often utilize frequency converters that supply power at a variety of low frequencies. A typical converter provides an output

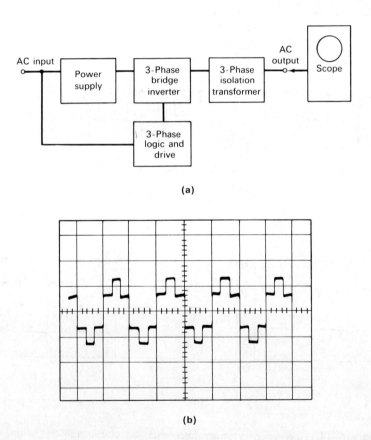

Figure 6–15 Three-phase industrial frequency converter: (a) block diagram; (b) normal output waveform.

SECT. 6-11 / FERRORESONANCE AND NEGATIVE IMPEDANCE 175

frequency from 380 to 1250 Hz at a power capability of 750 watts. A three-phase output is provided at 120 or 208 volts; the input is typically 120 volts at 60 Hz. Each phase normally has the output waveform depicted in Fig. 6–15. Observe that this is not a sine wave, but a step-type approximation. If this waveform is passed through a conventional 60-Hz power transformer, the higher harmonics will be reduced in amplitude, and the resulting waveform becomes a better approximation of a sine wave. In case of subnormal power output, an oscilloscope check may show that the output waveform is substantially distorted. In turn, the three-phase logic section falls under preliminary suspicion. If the logic waveform is normal, the drive transistors should be tested for collector-junction leakage or other defects.

6-11 FERRORESONANCE AND NEGATIVE IMPEDANCE

Considerable use is made of ferroresonance and its associated negative-impedance characteristic in various industrial-electronic devices. Any iron-cored coil that is driven over a substantial portion of its magnetic characteristic responds as a nonlinear inductor. In turn, its inductance value decreases as the current flow increases. If the nonlinear inductor is employed in a series LCR circuit, its resonant frequency will shift as the current value varies. This frequency shift is the operative feature of all ferroresonant circuits. With reference to the series LCR circuit depicted in Fig. 6–16, inductor L operates as a saturable reactor. At small current values, the circuit resonates at a frequency somewhat lower than that of the input voltage. As the current value I increases, voltage E_{LC} applied to the vertical-input terminal of the oscilloscope also increases. However, L soon starts to decrease in value, causing the circuit to resonate nearer the input frequency. In turn, the voltage drops across L and C becomes more nearly equal and are, of course, 180 deg apart in phase. Thus, a suitable increase of current causes E_{LC} to pass through a maximum value at A, and then to pass through a minimum value at B.

Point B occurs at the series-resonant frequency in Fig. 6–16. Note that E_{LC} does not fall completely to zero, owing to the winding resistance of L. Next, as the current value increases further, E_{LC} rises once more. Since the circuit passes through resonance, it follows that its reactance is inductive for small current values, and becomes capacitive for high current values. In other words, the current leads the applied voltage at first, and then lags as the current value increases. Over the interval from

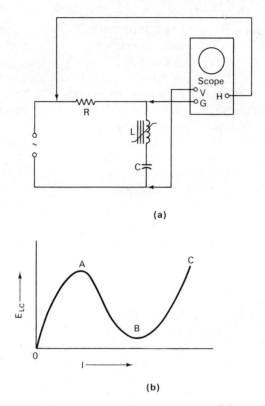

Figure 6–16 Basic ferroresonance arrangement: (a) configuration; (b) voltage-current characteristic.

A to B, an increase in current is accompanied by a decrease in voltage E_{LC}. Accordingly, this voltage-current characteristic exhibits a *negative impedance* over the interval from A to B. On the other hand, a positive impedance is exhibited from zero to A, and from B to C. A typical application of this characteristic is found in an automatic line-voltage regulating transformer.

6-12 MAGNETIC AMPLIFIER WAVEFORMS

Magnetic amplifiers are utilized in various types of industrial-electronics equipment. A magnetic amplifier is characterized by high efficiency (up to 90 percent), reliability, ruggedness, and zero warmup time. However, a magnetic amplifier cannot process low-level signals or high-frequency

SECT. 6-12 / MAGNETIC AMPLIFIER WAVEFORMS

signals. There is also a little time delay in response, and the output waveform is not a precise replica of the input waveform. In its appropriate area of application, the magnetic amplifier has no close competitor. A basic half-wave magnetic amplifier configuration with key waveforms is shown in Fig. 6–17.

A magnetic amplifier consists of a magnetic core made from a

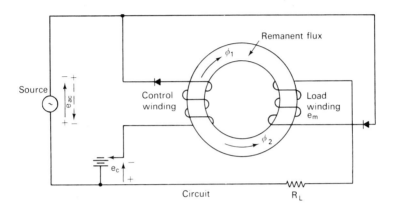

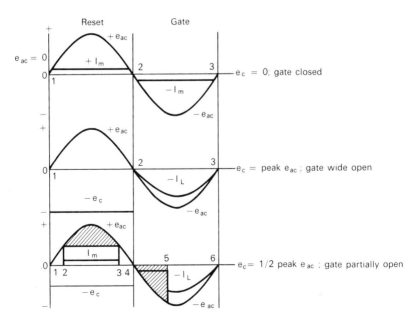

Figure 6–17 Basic half-wave magnetic amplifier.

special square-loop material upon which two windings are placed. Note that a square-loop magnetic material goes into saturation suddenly, compared with ordinary transformer iron. The load or gating winding is connected in series with a rectifier, the load, and an AC power supply. Next, the control winding is connected in series with a rectifier, the control-signal source, and the same AC source as the gating winding. Both windings have the same number of turns. Basically, the magnetic amplifier acts like an electrically operated contactor that turns on the load circuit periodically. A control voltage applied to the closing circuit of the contactor closes the contactor, which completes a circuit to the load. This action can be repeated periodically, for example, by introducing an AC control voltage in series with a half-wave rectifier and the contactor closing coil.

Note that the action of the control winding in a magnetic amplifier can be compared to that of the closing coil of a contactor. Conversely, the action of the load winding can be compared to that of the contactor's main contacts. This latter action consists of introducing a high impedance (main contacts open) for a controlled portion of each half-cycle and then removing this impedance (main contacts closed) and allowing the current to flow through the load during the remaining portion of the half-cycle. Rectifiers are placed in the load and control circuits to prevent current flow in the control circuit during the gating half-cycle and in the load, or gating, circuit during the reset half-cycle. Observe that a magnetic amplifier is not an amplifier in the sense of a step-up transformer. Voltages generated by mutual inductance (transformer action) between the control and load windings occur in these windings, but they have only a minor effect on the magnetic amplifier operation. In normal operation, a small control voltage e_c gates a large current flow through the load R_L.

APPENDIX 1

RMS VALUES OF BASIC COMPLEX WAVEFORMS

Root mean square values of various basic complex waveforms are indicated in Fig. A1–1. Although the average value of a complex waveform can be found by a simple oscilloscope test, it is not possible to measure its rms value directly with an oscilloscope. Note that when a complex waveform is applied to the AC input of an oscilloscope, the displayed waveform distributes itself with respect to the resting level of the beam so that the positive-peak excursion of the waveform is above the resting level, and the negative-peak excursion is below the resting level. There is no simple relation between the average value and the rms value of a complex waveform. Note that intuition is sometimes misleading with respect to rms values. As an illustration, the rms value of a half-rectified sine wave is equal to $V/2$, but the rms value of a full-rectified sine wave is equal to $V/\sqrt{2}$, where V is the peak voltage of the waveform.

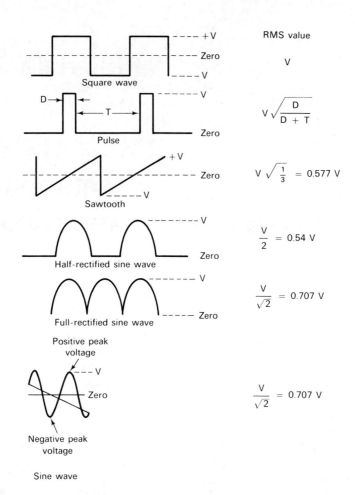

Figure A1-1 RMS values of some basic complex waveforms.

APPENDIX 2

BASIC FUNDAMENTAL AND SINGLE HARMONIC WAVEFORMS

It is helpful to the oscilloscope technican to recognize the basic fundamental and second-harmonic waveforms. Typical complex waveforms resulting from a mixture of a fundamental frequency with a second-harmonic, or third-harmonic, or fourth-harmonic, or fifth-harmonic frequency, are illustrated in Fig. A2–1. Observe that the waveshape can be changed considerably by shifting the phase of a harmonic with respect to its fundamental component. Odd harmonics always produce symmetrical positive and negative excursions. Even harmonics always produce asymmetrical positive and negative excursions. Note in Fig. A2–1(d) that the in-phase second harmonic does not produce a symmetrical resultant. In other words, the negative excursion is a mirror image of the positive excursion and is, therefore, asymmetrical. Note also that all of the waveforms shown in Fig. A2–1 are AC waveforms, and their average value is zero in each case. When a harmonic is shifted in phase, the resultant waveform still has an average value of zero. In other words, it is impossible to form a complex wave that has a DC component from any mixture of sine waves.

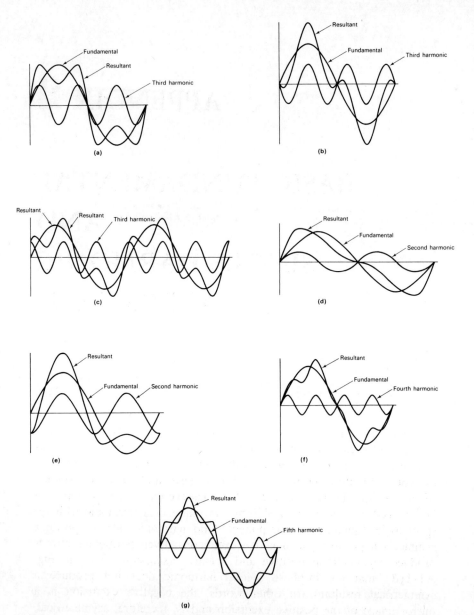

Figure A2-1 Basic fundamental and single harmonic waveforms: (a) fundamental with third harmonic in phase; (b) fundamental with third harmonic 180 deg out of phase; (c) fundamental with third harmonic 90 deg leading; (d) fundamental with second harmonic in phase; (e) fundamental with second harmonic 90 deg lagging; (f) fundamental with fourth harmonic in phase; (g) fundamental with fifth harmonic in phase.

APPENDIX 3

BASIC GEOMETRICAL CURVES

Waveforms that have the shapes of basic geometrical curves are encountered in electronic equipment. For example, the sine wave is widely utilized; a sine wave is a basic geometrical curve of the transcendental class. Parabolic waveforms are utilized in dynamic-convergence circuitry. As shown in Fig. A3–1, a parabola has approximately the same waveshape as a half-sine wave. A parabola is classified as a conic-section curve. Elliptical waveforms are exemplified by Lissajous figures, used to measure phase angles. An ellipse is one of the three conic-section curves, as shown in Fig. A3–2. Note that hyperbolic waveforms are employed in analog-computer systems. A diode characteristic can be approximated by a logarithmic curve, for most practical purposes. Basic diode and logarithmic waveforms are pictured in Fig. A3–3. The exponential waveform is another basic geometrical curve that is often encountered in electronic equipment. Fig. A3–4 depicts basic exponential curves.

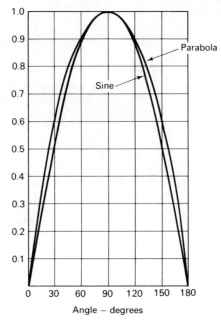

Figure A3–1 Comparison of parabola and half-sine waveforms.

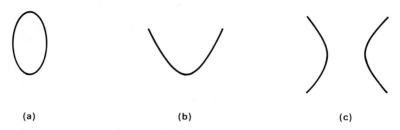

Figure A3–2 Comparison of the three conic sections: (a) ellipse; (b) parabola; (c) hyperbola.

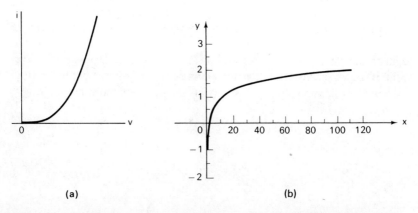

Figure A3–3 Diode and logarithmic curves: (a) diode voltage-current curve (standard aspect); (b) graph of $\log_{10} x$ (standard aspect).

APPENDIX 3 / BASIC GEOMETRICAL CURVES

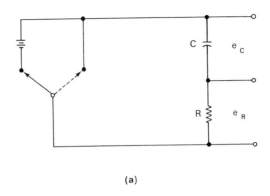

(a)

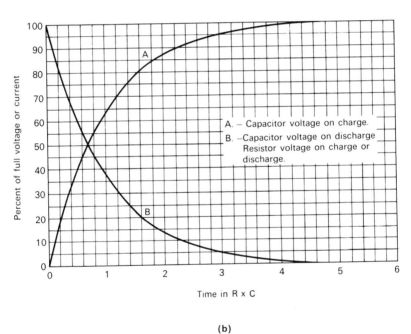

(b)

Figure A3-4 Rising and falling exponential curves: (a) basic circuit arrangement; (b) universal time-constant chart.

APPENDIX 4

COLOR-TV INPUT AND OUTPUT SYSTEM WAVEFORMS

Color-TV system waveforms start with red, green, and blue camera output waveforms at the transmitter. Typical camera output waveforms are shown in Fig. A4–1. The yellow signal is a mixture of red and green signals; the white signal is a mixture of red, green, and blue signals. These camera output waveforms are next processed into Y, R-Y, B-Y, and G-Y waveforms. At the color-TV receiver, these waveforms are converted into the original red, green, and blue waveforms. Finally, the red, green, and blue signals are applied to the corresponding electron guns in the color picture tube.

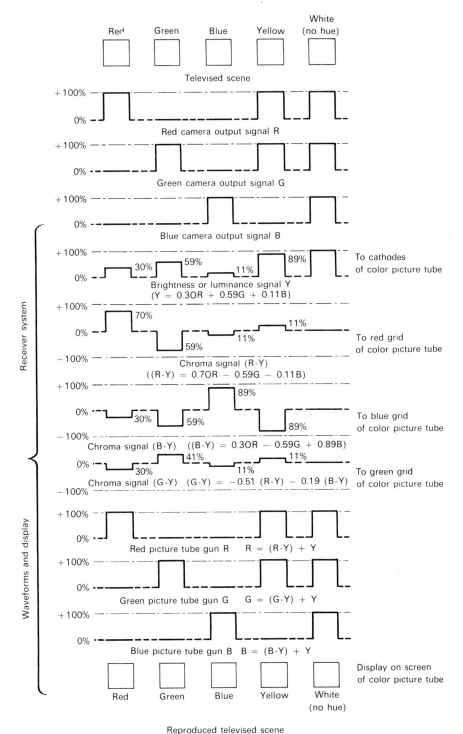

Figure A4-1 Color-TV system waveforms.

APPENDIX 5

I-Q COLOR WAVEFORM CHART

Color-TV transmission is accomplished on the I and Q chroma axes. The $\pm$I and $\pm$Q signals are called the transmission primaries. Their relations to the additive primary colors are shown in Fig. A5–1. Although color-TV transmission occurs on the I and Q axes, chroma demodulation may be accomplished on any pair of chroma axes. All standard color-TV receivers employ R-Y and B-Y chroma demodulators, or X and Z chroma demodulators. The X and Z axes are somewhat arbitrary, but are always chosen in the vicinity of the R-Y and B-Y axes. X and Z chroma demodulation is used in the interest of production economy. Phase relations of the primary color signals and the principal color-difference signals are shown in Fig. A5–2.

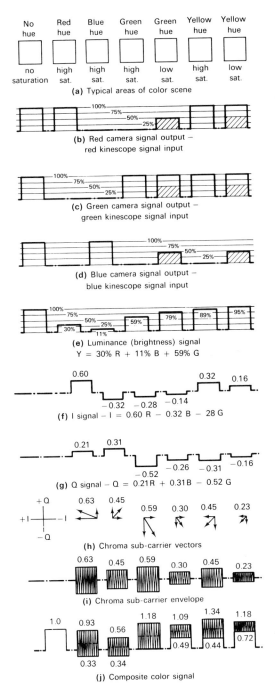

Figure A5–1 I-Q chroma signal relations to the additive primary colors.

APPENDIX 5 / I-Q COLOR WAVEFORM CHART

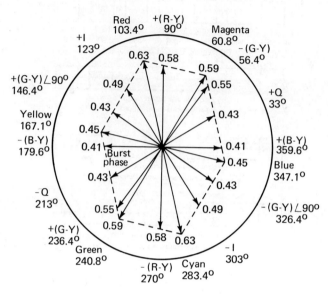

Figure A5–2 Phase relations of the primary color signals and the principal color-difference signals.

APPENDIX 6

VECTORSCOPE WAVEFORM DEVELOPMENT

TV technicians often utilize vectorgrams for an overall check of chroma demodulation and/or matrixing action. Vectorgrams are specialized Lissajous patterns obtained with a color-bar signal. In most cases, a keyed-rainbow signal is used (Fig. A6–1). An oscilloscope is connected at the outputs of the R-Y and B-Y demodulators, as shown in Fig. A6–2. Alternatively, the oscilloscope may be connected at the matrix outputs. Keyed-rainbow signal vector relationships are depicted in Fig. A6–3. When R-Y and B-Y signal inputs are applied to the oscilloscope, the ideal vectorgram development occurs as shown in Fig. A6–4. Because of blanking action, only ten petals appear in practice, as exemplified in Fig. A6–5; in other words, the eleventh and twelfth phase components are blanked in the receiver. Petals have rounded tops because of limited bandwidth in the chroma channel. Angular relations between successive petals may be more or less incorrect owing to demodulation phase errors. Petal waveshapes are somewhat distorted in practice due to residual irregularities in the demodulation process.

APPENDIX 6 / VECTORSCOPE WAVEFORMS

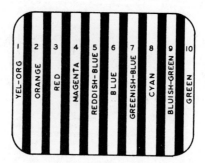

Figure A6–1 Display of a keyed-rainbow color-bar signal on the picture-tube screen.

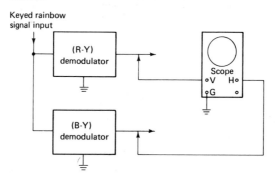

Figure A6–2 Typical vectorscope test setup.

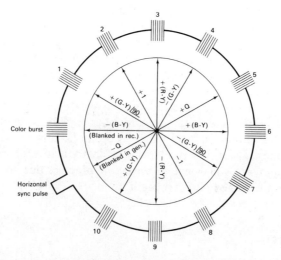

Figure A6–3 Keyed-rainbow signal vector relationships.

APPENDIX 6 / VECTORSCOPE WAVEFORMS

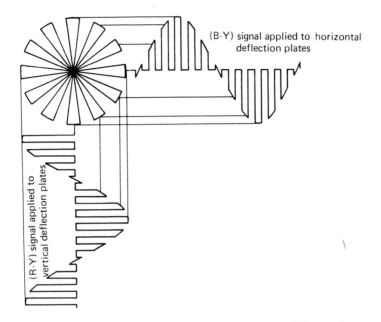

Figure A6–4 Ideal vectorgram development. (*Courtesy of* Sencore)

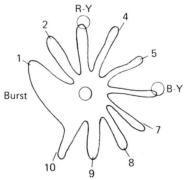

Figure A6–5 Typical vectorgram display for R-Y/B-Y demodulators. (*Courtesy of* Sencore)

APPENDIX 7

BALANCED MODULATOR WAVEFORMS

Three basic types of balanced-modulator arrangements are utilized in television technology. These configurations are called the video-balanced modulator, the carrier-balanced modulator, and the doubly-balanced modulator. As shown in Fig. A7–1, a video-balanced modulator changes carrier and video-input waveforms into a carrier-and-sideband waveform. In other words, the video-waveform is cancelled out, but the carrier-input waveform feeds through to the output. Next, the carrier-balanced modulator changes carrier and video-input waveforms into a video-and-sideband waveform. Thus, the carrier-input waveform is cancelled out, but the video-input waveform feeds through to the output. Finally, the doubly-balanced modulator changes carrier and video-input waveforms into a sideband output waveform. That is, both the video-input waveform and the carrier-input waveform are cancelled out, leaving only the sideband waveform in the output.

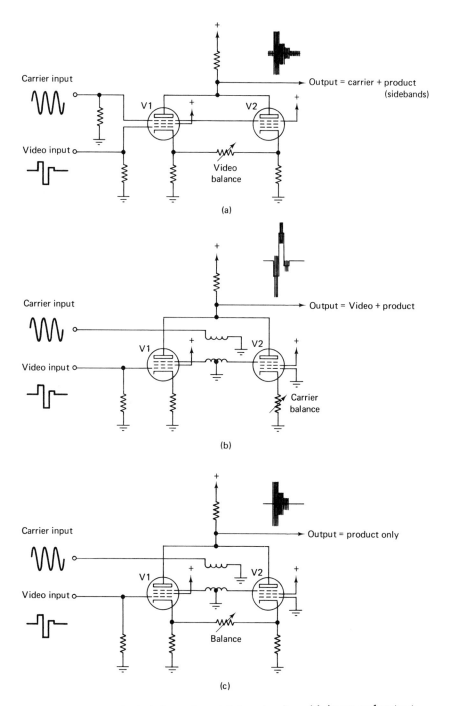

Figure A7–1 Three basic balanced-modulator circuits, with input and output waveforms: (a) video-balanced modulator; (b) carrier-balanced modulator; (c) doubly-balanced modulator.

INDEX

Abnormal bandwidth, 132
Absolute values, 27
Absorption markers, 137
AFPC, 143
AGC, 101
Alignment, 83
AM detector, 80
Amplifier
 magnetic, 176
 sensitivity, 62
Analog computers, 2
Antilog operation, 7
Aspect, curve, 93
Audio
 information, 121
 oscillator, 46
 sweep, 41
Average power, 48

Balanced modulator, 193
Bandpass
 filter, 19
 and matrix, 83

Bandwidth, 44
Baseline
 curvature, 83, 146
 irregularities, 24
Bass controls, 42
Beat marker, 97
Bias box, 109
Binaural, 70
Blanking, 66, 95
 pulse, 119
Bridge, switching, 81
Bridged equalizer, 33
Burst waveform, 25

Calibrated time base, 58
Chroma demodulator, 143
Circuit loading, 99
Compatibility, 70
Compensation, 115
Complementary symmetry, 63
Complex number, 31
Composite audio, 68
Compression, 41

Conventionalized waveforms, 25
Corner rounding, 60, 104
Counter electromotive force, 19
Crossover distortion, 56
Cube, sine-wave, 11

DC component, 6
Decoder, 67, 72, 85
Decoding process, 3
Delay line, 32
Delta sound, 121
Demodulation capability, 99
Demodulator probe, 99
Detector, traveling, 99
Device isolation, 19
Differentiation, 16
Digital pulses, 24
Distortion
  crossover, 56
  meter, 53
  products, 53
  square-wave, 59
Distributed capacitance, 111

Encoding, 72
Equalizer, 33
Equalizing pulses, 104
Equivalent inductive response, 19
Evaluation, distortion, 51
Evolution, waveform, 7
Expanded burst, 137
Expansion, 41

Ferroresonance, 175
Finite rise time, 24
FM stereo, 73
Frequency
  counter, digital, 143
  distortion, 39
  doubling, 6
  modulation, 70, 120
  spectrum, 74
Front end, 93

Gating winding, 178
Generated waveforms
  internal, 91
  nonuniform, 95

Generator
  stereo, 71
  test-pattern, 104
Graticule, 66
G–Y waveform, 27

Half-power bandwidth, 44
Harmonic distortion, 39
  percentage, 46
HF component, 81
High fidelity, 39
High-pass filter, 65
Horizontal sync pulse, 99
Hum and noise, 52

Ideal waveforms, 21
IF alignment, 107
Imaginary
  current, 31
  number, 31
  power, 31
Impedance, negative, 176
Information, sync, 105
Instantaneous power, 47
Integrating counter, 161
Integration, 16
Integrator, op-amp, 7
Interference, station, 107
Intermodulation distortion, 39, 63
Inverter, polarity, 75
Involution, 7
I–Q, 188
Iron third harmonic, 12
Isolating resistor, 109

J (operator), 31

Key
  checkpoints, 85
  waveforms, 25, 84
Keyed rainbow, 145

Laboratory oscilloscope, 99
Lag, 28
Lead, 28
Leading, circuit, 99
Leading edge, 57
Limiter, 85

INDEX

Line crawl, 134
Lissajous patterns, 44
Load impedance, 49
Log operation, 7
Logic section, 175
Loudness control, 42
L–R signal, 73
L + R signal, 73
L signal, 68, 74

Malfunction, instrument, 95
Marker
  adder, 99
  injection, 98
  pip, 88
  waveshaper, 99
Mathematical integral, 7
Maximum rated power, 44
Misalignment, 83
Modified exponential, 16
Modulation
  balanced, 193
  downward, 122
Monophonic, 70
Multiplex, 67, 72
Music power, 47
Musical tone, 47

Negative
  excursion, 27
  impedance, 176
Noise
  "blue," 85
  interference, 127
  "red," 85
  waveforms, 85
  "white," 85
Nonlinear
  amplification, 39
  mixing, 13
  operation, 11
NTSC, 132

Ohmic value, 61
Op-amp subtracter, 10
Operational amplifiers, 4
Oscillation, parasitic, 52

Oscilloscope
  storage type, 41
  triggered-sweep, 35
Output signal, composite, 69
Overdamped circuit, 170
Overload, 101
Overshoot, 24, 116

Parabolic waveform, 7
Parasitic oscillation, 52
Peak power, 47, 61
Peak-to-peak voltage, 27
Phase relations, 27
Picture detail, 104
Pilot subcarrier, 77
Positive-going, 21
Power
  amplifier, 42
  bandwidth, 44
Probe, demodulator, 99, 113
Product waveforms, 1
Pulse waveforms, 60

Quadrupled peak power, 48

Ramp, 120
Rated power, 46
Reactive power, 31
Real
  power, 31
  waveforms, 21
Reconstitution, 78
Resultant waveform, 55
Return trace, 95
Ringing, 112
Rise time, 58
RMS
  power, 46
  values, 179
Root extraction, 11
R signal, 68, 74
Rumble filter, 42

Scratch filter, 42
Selectivity, 101
Sensitivity, amplifier, 62
Separation, 67

Serrations, 119
Sidebands, 12
Sine-squared waveform, 7
Square
  law, 47
  wave, 52
Steady state, 1
  distortion, 51
Stereophonic, 70
Storage oscilloscope, 41
Subcarrier, 77
  pilot, 77
  suppression, 77
Summing probe, 2
Synthesis, square wave, 58

Three-section integrator, 18
Time base, 57
Tone waveform, 40
Transient, 1
  distortion, 52
Traveling detector, 99
Treble controls, 42
Triggered sweep, 35, 57
True power, 31
TV sync waveforms, 23
Two-section integrator, 18
Two-tone signal, 63

Unbalanced output, 63
Underdamping, 170
Undershoot, 19
Universal time constant, 7
Upward modulation, 122

VAR, 31
Vector diagram, 29
Vectorscope, 67
Vertical sync pulse, 99
VHF reception, 99
Video
  amplifier, 109
  detector, 100
VIR, 153
VITS, 153
Volt-amperes reactive, 31
Voltage
  gain, 62
  negative-peak, 27
  peak-to-peak, 27
  positive-peak, 27
  ratio, 69
  squared waveform, 47
  waveforms, 91
VSM, 136

Wattless power, 31
Waveform
  analysis, 37
  current, 91
  filtering, 3
  multiplier, 4
  number relations, 25
  relations, 22
  rise, 18
  signal, 91
  sine-squared, 7
  spectrum, 91
  vowel sound, 41
Waveshaping amplifier, 14, 117
White
  highlights, 101
  saturation, 109

X and Z axes, 188

Y amplifier, 127

Z matching, 122
Z, negative, 176
Zero
  level, 27, 96
  rise time, 24